RESOURCES

RESOURCES

FOR SCIENCE LITERACY

PROFESSIONAL DEVELOPMENT

PROJECT 2061

AMERICAN ASSOCIATION FOR THE ADVANCEMENT OF SCIENCE

OXFORD UNIVERSITY PRESS

NEW YORK   OXFORD

1997

OXFORD UNIVERSITY PRESS

Oxford   New York

Athens   Auckland   Bangkok   Bogota   Bombay
Buenos Aires   Calcutta   Cape Town   Dar es Salaam
Delhi   Florence   Hong Kong   Istanbul
Karachi   Kuala Lumpur   Madras   Madrid
Melbourne   Mexico City   Nairobi   Paris
Singapore   Taipei   Tokyo   Toronto

AND ASSOCIATED COMPANIES IN

Berlin   Ibadan

Library of Congress Cataloging-in-Publication Data

Project 2061 (American Association for the Advancement of Science)
Resources for science literacy : professional development / Project 2061,
American Association for the Advancement of Science.
p.   cm.
Includes bibliographical references.
ISBN 0-19-510873-6 (bk & disk)
1. Science—Study and teaching—United States. 2. Project 2061 (American Association for the Advancement of Science)
Q183.3.A1P76   1997
507'.1'273—dc21
96-49376
CIP

The *Resources for Science Literacy: Professional Development* CD-ROM was developed using Micromedia Authorware.

Authorware is a registered trademark of Macromedia, Inc.
Macintosh is a trademark of Apple Computer, Inc.
Windows is a trademark of Microsoft Corporation.
Acrobat is a trademark of Adobe Systems, Inc.

1 3 5 7 9 8 6 4 2
Printed in the United States of America
on acid-free paper

TABLE OF CONTENTS

Founded in 1848, the **American Association for the Advancement of Science** (AAAS) is the world's largest federation of scientific and engineering societies, with nearly 300 affiliate organizations. In addition, AAAS counts more than 140,000 scientists, engineers, science educators, policy makers, and interested citizens among its individual members, making it the largest general scientific organization in the world. The Association's goals are to further the work of scientists, facilitate cooperation among them, foster scientific freedom and responsibility, improve the effectiveness of science in the promotion of human welfare, advance education in science, and increase public understanding and appreciation of the importance and promise of the methods of science in human progress.

Project 2061 is a long-term initiative of AAAS to reform K-12 education in natural and social science, mathematics, and technology. Begun in 1985, Project 2061 is developing a comprehensive set of science education reform tools to help educators make science literacy a reality for all American students. With the publication in 1989 of *Science for All Americans*, Project 2061 defined science literacy in terms of the knowledge and skills that all high school graduates need. To provide more specific guidance on how students should progress toward science literacy, *Benchmarks for Science Literacy* (1993) describes what students should know and be able to do in science, mathematics, and technology by the end of grades 2, 5, 8, and 12. Project 2061 continues to develop a variety of print and electronic tools and provides workshops and other professional development services to science educators nationwide.

The AAAS wishes to express its gratitude to the following for their generous support of Project 2061:

CARNEGIE CORPORATION OF NEW YORK
JOHN D. AND CATHERINE T. MACARTHUR FOUNDATION
ANDREW W. MELLON FOUNDATION
ROBERT N. NOYCE FOUNDATION
THE PEW CHARITABLE TRUSTS
NATIONAL SCIENCE FOUNDATION
U.S. DEPARTMENT OF EDUCATION

PREFACE

Over the past decade Project 2061 has developed science literacy goals for education reform. With the publication of *Science for All Americans* in 1989 and *Benchmarks for Science Literacy* in 1993—along with the recent publication of the National Research Council's *National Science Education Standards*—there is now a strong consensus on what students should know and be able to do in science, mathematics, and technology by the time they graduate from high school.

In many ways, defining and setting these goals has been the easy part of reform. The more difficult tasks still remain: understanding the implications of goal-directed reform, implementing reform strategies with those goals in mind, and making systemwide changes that are consistent with those goals.

Resources for Science Literacy: Professional Development is a first step toward that change. It brings together the collective expertise of Project 2061 staff and that of teachers and professional development experts from around the country. We have set out to help solve some of the most immediate problems educators face as they begin to use science literacy goals in their classrooms and their schools—to rethink the way they develop curricula, select materials and activities, design instruction, and plan for assessment. *Resources for Science Literacy: Professional Development* is a practical, multi-purpose tool that educators can use for these and many other purposes and in a variety of ways.

Throughout this volume, you will see requests for your ideas, suggestions, and comments. We urge you to respond. Feedback from you is a vital part of Project 2061's effort to create sound, field-tested tools that educators can use to reform America's schools. We invite you to join us as partners in that effort.

F. JAMES RUTHERFORD
Director, Project 2061

NATIONAL COUNCIL ON SCIENCE AND TECHNOLOGY EDUCATION

CHAIRMAN

Donald Langenberg   *Chancellor*, University of Maryland Systems

Raul Alvarado, Jr.   *Small Business Office Space Station Division*, McDonnell-Douglas Corporation

William O. Baker   *Retired Chairman of the Board*, AT&T Bell Telephone Laboratories

Catherine Belter   *Chair, PTA Education Commission*, The National PTA

Diane J. Briars   *Director*, Division of Mathematics, Connelley Technical School Support Services, Pittsburgh Public Schools

Patricia L. Chavez   *Director of Corporate Relations and Advancement*, The University of New Mexico

Marvin Druger   *Chairman and Professor*, Department of Science Teaching, Syracuse University

Joan Duea   *Professor of Education*, University of Northern Iowa

Bernard Farges   *Mathematics Teacher*, San Francisco Unified School District

Stuart Feldman   *Department Group Manager*, Internet Applications and Services, T.J. Watson Research Center, IBM

Linda Froschauer   *Science Department Chair Person*, Weston Middle School, Connecticut

Patsy D. Garriott   *Education Initiatives Representative*, Eastman Chemical Company

Gregory A. Jackson   *Director of Academic Computing*, Massachusetts Institute of Technology

Cherry H. Jacobus   *Vice President*, Marketing, Goodwill Industries

Fred Johnson   *Assistant Superintendent for Instruction*, Shelby County Board of Education, Tennessee

Roberts T. Jones   *President and Chief Executive Officer*, National Alliance of Business

David Kennedy   *State Science Supervisor*, Washington

George Kourpias   *President*, International Association of Machinists and Aerospace Workers

Kenneth R. Manning   *Professor of the History of Science*, Massachusetts Institute of Technology

Sue Matthews   *Science Teacher*, Elbert County School District, Georgia

Jose F. Mendez   *President*, Ana G. Mendez University System

George Nelson   *Associate Vice Provost for Research*, University of Washington

Freda Nicholson   *Executive Director*, Science Museums of Charlotte, Inc.

James R. Oglesby   University of Missouri, Columbia

Gilbert S. Omenn   *Dean*, School of Public Health and Community Medicine, University of Washington

Lee Etta Powell   *Professor of Educational Administration*, George Washington University

Vincent E. Reed   *Vice President*, Communications, The Washington Post

Thomas Romberg   *Director*, National Center for Research in Mathematical Sciences Education

Mary Budd Rowe   *Professor of Science Education*, Stanford University

David Sanchez   *Professor of Mathematics*, Texas A&M University System

Albert Shanker   *President*, American Federation of Teachers

Benjamin S. Shen   *Reese W. Flower Professor of Astronomy and Astrophysics*, University of Pennsylvania

Claibourne D. Smith   *Vice President*, Technical-Professional Development, DuPont Company

Gloria Takahashi   *Teacher*, Science Department, La Habra High School, La Habra, California

Walter B. Waetjen   *President Emeritus*, Cleveland State University

William Winter   *Attorney-at-Law*, Watkins Ludlam & Stennis

Terry Wyatt   *Science Teacher*, Roy C. Start High School, Toledo, Ohio

John Zola   *Teacher*, Social Sciences, The New Vista High School, Boulder, Colorado

EX-OFFICIO MEMBERS

Francisco J. Ayala   *Donald Bren Professor of Biological Sciences*, University of California, Irvine

F. James Rutherford   *Director*, Project 2061, *Chief Education Officer*, American Association for the Advancement of Science

WINSLOW HOMER,
Blackboard, 1877.

INTRODUCTION

Ten years ago, Project 2061 of the American Association for the Advancement of Science (AAAS) set out to define what it means to be science literate. Although educators, political leaders, parents, and community members agreed on the importance of science literacy for citizens of the modern world, no one had clearly and explicitly addressed the question: "What should a science-literate adult know and be able to do in science, mathematics, and technology?"

PROJECT 2061 DEFINES SCIENCE LITERACY

Drawing on the work of several scientific panels convened by AAAS to develop an answer, Project 2061 took on the task of setting out for the nation the knowledge, skills, and habits of mind that all citizens need to live interesting, responsible, and productive lives in a culture shaped by science and technology. The result of that effort was *Science for All Americans*, published by Project 2061 in 1989. The next step was to map out how student understanding should progress from the relatively simple ideas intelligible to kindergartners toward the more sophisticated knowledge recommended for high school graduates in *Science for All Americans*. In its 1993 report *Benchmarks for Science Literacy*, Project 2061 did just that, recommending a coherent set of learning goals in science, mathematics, and technology for students in grades 2, 5, 8, and 12. Together, these two publications have helped to establish science literacy as the basis for other national, state, and local reform efforts and have greatly influenced the development of the content portion of *National Science Education Standards* released by the National Research Council in 1996.

With this growing national consensus on the content of the science curriculum now underway, Project 2061 has turned its attention to helping educators implement reforms at many levels. One example is this CD-ROM/ print tool, *Resources for Science Literacy: Professional Development*. Together with *Resources for Science Literacy: Curriculum Materials*, which is now being developed, this new tool will help educators to:

- enhance their own knowledge of science, mathematics, and technology;
- gain a better understanding of science literacy goals; and
- make well-informed decisions about curriculum resources and materials.

Another reform tool under development is *Designs for Science Literacy*, which will help educators take a systematic approach to the design of K-12 curricula that target science literacy goals such as those presented in *Science for All Americans* and *Benchmarks*. And because the education system as a whole must change to accommodate many curriculum reforms, Project 2061 has commissioned a dozen expert committees to examine various aspects of the system, such as school organization, equity, teacher education, assessment, and more. Their findings will be integrated into *Blueprints for Reform*, a report to the nation identifying the organizational, structural, and policy changes that will be required.

UNDERSTANDING SCIENCE LITERACY GOALS

In *Science for All Americans*, the notion of science literacy is construed broadly. It includes understandings about the natural and the social sciences, mathematics, technology, and the connections among them. This inclusiveness reflects Project 2061's conviction that understanding any one of these areas will require understanding the others. *Science for All Americans* further extends the scope of science literacy to include knowledge about the nature of the scientific enterprise itself, historical episodes of exceptional significance in the development of science, and themes that cut across science, mathematics, and technology.

Few teachers can be expected to understand all of the topics in *Science for All Americans*, how they interconnect, and what they imply for science, mathematics, and technology education at all grade levels. Even teachers who have a firm grounding in the science content and are confident they can put *Science for All Americans* to good use in planning curricula, often find other obstacles—inadequate curriculum materials and a lack of research about how to teach some of the topics in *Science for All Americans*, for example. Project 2061 has created *Resources for Science Literacy: Professional Development* to help educators overcome these difficulties.

Resources for Science Literacy: Professional Development

This new CD-ROM/print tool was developed in response to the many requests Project 2061 receives from educators who want to know more about understanding and using *Science for All Americans* and *Benchmarks for Science Literacy*. It is also designed to help teachers fill in gaps in their own knowledge of science, mathematics, and technology and their interconnections. It will enable teachers to increase their understanding of science literacy and their familiarity with the ideas and skills that students of various ages need to develop on their way to science literacy.

Resources for Science Literacy: Professional Development offers a carefully selected collection of references, workshop activities, research, analyses, and course plans that illustrates many aspects of science literacy and its application to K-12 science education. It can be used by teachers for self-guided study of *Science for All Americans* and *Benchmarks* or for developing pre-service and in-service education programs.

Multiple links allow users to search for resources that relate to specific topics presented in *Science for All Americans*. For example, users can create a customized list of recommended trade books dealing with scientific inquiry (found in *Science for All Americans* Chapter 1: The Nature of Science), then review the specific learning goals related to that topic recommended in *Benchmarks for Science Literacy* and *National Science Education Standards*, go on to explore the available cognitive research on student learning of concepts related to scientific inquiry, and, finally, examine the college course syllabi for some guidance on developing a systematic approach to learning more about the topic. With one exception (noted below), all of the databases included on the *Resources for Science Literacy: Professional Development* CD-ROM are linked by hypertext to chapters and sections in *Science for All Americans* and to one another.

The CD-ROM includes six components:

Science for All Americans. The book's full text is accessible on the CD-ROM and is linked to all of the other components (except the Project 2061 *Workshop Guide*). These links enable users to identify resources that are relevant to specific chapters and sections in *Science for All Americans*.

Science Trade Books. More than 120 books for general readers dealing with all areas of science, technology, and mathematics are cited with full bibliographic information, reviews, and other descriptive data. Each book is linked to specific *Science for All Americans* chapters and sections so that users can compile a reading list around *Science for All Americans* topics.

Cognitive Research. An introduction to cognitive research literature sheds light on the ability of students of various ages to understand many of the topics in *Science for All Americans* and *Benchmarks for Science Literacy*. In addition, *Benchmarks'* Chapter 15: The Research Base—and its accompanying bibliography of more than 300 references—is included on the CD-ROM.

College Courses. Descriptions of 15 undergraduate college courses suggest some guidelines and examples for designing syllabi to teach college students particular concepts from *Science for All Americans*. Links take the user directly to the relevant chapters and sections of *Science for All Americans*.

Comparisons of *Benchmarks* to National Standards. Included here are analyses of how *Benchmarks for Science Literacy* relates to three sets of national content standards—the National Research Council's *National Science Education Standards*, the National Council of Teachers of Mathematics' *Curriculum and Evaluation Standards for School Mathematics*, and the National Council for the Social Studies' *Curriculum Standards for Social Studies*.

Project 2061 Workshop Guide. Developed and field-tested by Project 2061 staff, teachers from Project 2061's six School-District Centers, and education consultants, the *Guide* includes a variety of presentations, activities, and supplementary materials that can be used to design Project 2061 workshops or as a tutorial on *Science for All Americans* and *Benchmarks*. Users can access other *Professional Development* components from a number of points within the *Guide*.

This companion volume presents examples from each of the six *Professional Development* components. It also provides background information on how and why each component was developed and suggestions for its use.

The release of the *Professional Development* CD-ROM will be followed by *Resources for Science Literacy: Curriculum Materials*. It too will be available as a CD-ROM with a companion print volume. Eventually, both *Professional Development* and *Curriculum Materials* will be revised, expanded, and merged into a single *Resources for Science Literacy* CD-ROM that will be periodically updated.

A WORK IN PROGRESS

Resources for Science Literacy: Professional Development is an evolving tool. To create this first release, Project 2061 consulted numerous teachers and other experts in science, mathematics, and technology education. Yet this compilation of materials is by no means exhaustive. Project 2061 invites you and your colleagues to recommend additional science trade books, research studies, or college courses that you have encountered and would like to share. Your suggestions will be considered for the next update to *Resources for Science Literacy: Professional Development*.

Send suggestions to:
Project 2061
American Association for the Advancement of Science
1333 H Street, NW
P.O. Box 34446
Washington, D.C. 20005
FAX: 202/842-5196
Electronic Mail: project2061@aaas.org (please identify subject as "feedback")

RESOURCES

Chapter **1** ABOUT *SCIENCE FOR ALL AMERICANS*

Science for All Americans (*SFAA*) is the foundation of Project 2061's strategy to reform science education. It describes what all students should know and be able to do in science, mathematics, and technology by the time they graduate high school. Based on the work of the National Council on Science and Technology Education—a distinguished group of scientists and educators appointed by the American Association for the Advancement of Science—*SFAA* recommends the knowledge and habits of mind that will enable all citizens to make sense of how the world works, to think critically and independently, and to lead interesting, responsible, and productive lives in a culture increasingly shaped by science and technology. In addition, *Science for All Americans* lays out some principles of effective learning and teaching. Introductory text for each of the *Science for All Americans* chapters is presented below.

The full text of *Science for All Americans* is accessible on the *Resources for Science Literacy: Professional Development* CD-ROM. The volume's science literacy goals are linked directly to several of the other components on the disk, i.e., Science Trade Books, Cognitive Research, and College Courses.

CONTENTS OF
SCIENCE FOR ALL AMERICANS
Chapter 1 / **The Nature of Science**
The Scientific World View
Scientific Inquiry
The Scientific Enterprise

Over the course of human history, people have developed many interconnected and validated ideas about the physical, biological, psychological, and social worlds. Those ideas have enabled successive generations to achieve an increasingly comprehensive and reliable understanding of the human species and its environment. The means used to develop these ideas are particular ways of observing, thinking, experimenting, and validating. These ways represent a fundamental aspect of the nature of science and reflect how science tends to differ from other modes of knowing.

It is the union of science, mathematics, and technology that forms the scientific endeavor and that makes it so successful. Although each of these human enterprises has a character and history of its own, each is dependent on and reinforces the others. Accordingly, the first three chapters of recommendations draw portraits of science, mathematics, and technology that emphasize their roles in the scientific endeavor and reveal some of the similarities and connections among them.

This chapter lays out recommendations for what knowledge of the way science works is requisite for scientific literacy. The chapter focuses on three principal subjects: the scientific world view, scientific methods of inquiry, and the nature of the scientific enterprise. Chapters 2 and 3 consider ways in which mathematics and technology differ from science in general. Chapters 4 through 9 present views of the world as depicted by current science; Chapter 10: Historical Perspectives, covers key episodes in the development of science; and Chapter 11: Common Themes, pulls together ideas that cut across all these views of the world.

Chapter 2 / **The Nature of Mathematics**
Patterns and Relationships
Mathematics, Science, and Technology
Mathematical Inquiry

Mathematics relies on both logic and creativity, and it is pursued both for a variety of practical purposes and for its intrinsic interest. For some people, and not only professional mathematicians, the essence of mathematics lies in its beauty and its intellectual challenge. For others, including many scientists and engineers, the chief value of mathematics is how it applies to their own work. Because mathematics plays such a central role in modern culture, some basic understanding of the nature of mathematics is requisite for scientific literacy. To achieve this, students need to perceive mathematics as part of the scientific endeavor, comprehend the nature of mathematical thinking, and become familiar with key mathematical ideas and skills.

This chapter focuses on mathematics as part of the scientific endeavor and then on mathematics as a process, or way of thinking. Recommendations related to mathematical ideas are presented in Chapter 9: The Mathematical World, and those on mathematical skills are included in Chapter 12: Habits of Mind.

Chapter 3 / **The Nature of Technology**
Technology and Science
Design and Systems
Issues in Technology

As long as there have been people, there has been technology. Indeed, the techniques of shaping tools are taken as the chief evidence of the beginning of human culture. On the whole, technology has been a powerful force in the development of civilization, all the more so as its link with science has been forged. Technology—like language, ritual, values, commerce, and the arts—is an intrinsic part of a cultural system and it both shapes and reflects the system's values. In today's world, technology is a complex social enterprise that includes not only research, design, and crafts but also finance, manufacturing, management, labor, marketing, and maintenance.

In the broadest sense, technology extends our abilities to change the world: to cut, shape, or put together materials; to move things from one place to another; to reach farther with our hands, voices, and senses. We use technology to try to change the world to suit us better. The changes may relate to survival needs such as food, shelter, or defense, or they may relate to human aspirations such as knowledge, art, or control. But the results of changing the world are often complicated and unpredictable. They can include unexpected benefits, unexpected costs, and unexpected risks—any of which may fall on different social groups at different times. Anticipating the effects of technology is therefore as important as advancing its capabilities.

This chapter presents recommendations on what knowledge about the nature of technology is required for scientific literacy and emphasizes ways of thinking about technology that can contribute to using it wisely. The ideas are sorted into three sections: the connection of science and technology, the principles of technology itself, and the connection of technology and society. Chapter 8: The Designed World, presents principles relevant to some of the key technologies of today's world. Chapter 10: Historical Perspectives, includes a discussion of the Industrial Revolution. Chapter 12: Habits of Mind, includes some skills relevant to participating in a technological world.

Chapter 4 / **The Physical Setting**

The Universe
The Earth
Processes That Shape the Earth
Structure of Matter
Energy Transformations
Motion
Forces of Nature

Humans have never lost interest in trying to find out how the universe is put together, how it works, and where they fit in the cosmic scheme of things. The development of our understanding of the architecture of the universe is surely not complete, but we have made great progress. Given a universe that is made up of distances too vast to reach and of parti-

cles too small to see and too numerous to count, it is a tribute to human intelligence that we have made as much progress as we have in accounting for how things fit together. All humans should participate in the pleasure of coming to know their universe better.

This chapter consists of recommendations for basic knowledge about the overall structure of the universe and the physical principles on which it seems to run, with emphasis on the earth and the solar system. The chapter focuses on two principal subjects: the structure of the universe and the major processes that have shaped the planet earth, and the concepts with which science describes the physical world in general —organized for convenience under the headings of matter, energy, motion, and forces.

Chapter 5 / **The Living Environment**

Diversity of Life
Heredity
Cells
Interdependence of Life
Flow of Matter and Energy
Evolution of Life

People have long been curious about living things— how many different species there are, what they are like, where they live, how they relate to each other, and how they behave. Scientists seek to answer these questions and many more about the organisms that inhabit the earth. In particular, they try to develop the concepts, principles, and theories that enable people to understand the living environment better.

Living organisms are made of the same components as all other matter, involve the same kind of transformations of energy, and move using the same basic kinds of forces. Thus, all of the physical principles discussed in Chapter 4: The Physical Setting, apply to life as well as to stars, raindrops, and television sets. But living organisms also have characteristics that can be understood best through the application of other principles.

This chapter offers recommendations on basic knowledge about how living things function and how they interact with one another and their environment. The chapter focuses on six major subjects: the diversity of life, as reflected in the biological characteristics of the earth's organisms; the transfer of heritable characteristics from one generation to the next; the structure and functioning of cells, the basic building blocks of all organisms; the interdependence of all organisms and their environment; the flow of matter and energy through the grand-scale cycles of life; and how biological evolution explains the similarity and diversity of life.

Chapter 6 / **The Human Organism**
Human Identity
Human Development
Basic Functions
Learning
Physical Health
Mental Health
As similar as we humans are in many ways to other species, we are unique among the earth's life forms in our ability to use language and thought. Having evolved a large and complex brain, our species has

a facility to think, imagine, create, and learn from experience that far exceeds that of any other species. We have used this ability to create technologies and literary and artistic works on a vast scale, and to develop a scientific understanding of ourselves and the world.

We are also unique in our profound curiosity about ourselves: How are we put together physically? How were we formed? How do we relate biologically to other life forms and to our ancestors? How are we as individuals like or unlike other humans? How can we stay healthy? Much of the scientific endeavor focuses on such questions.

This chapter presents recommendations for what scientifically literate people should know about themselves as a species. Such knowledge provides a basis for increased awareness of both self and society. The chapter focuses on six major aspects of the human organism: human identity, human development, the basic functions of the body, learning, physical health, and mental health. The recommendations on physical and mental health are included because they help relate the scientific understanding of the human organism to a major area of concern—personal well-being—common to all humans.

Chapter 7 / **Human Society**

Cultural Effects on Behavior
Group Behavior
Social Change
Social Trade-Offs
Political and Economic Systems
Social Conflict
Global Interdependence

As a species, we are social beings who live out our lives in the company of other humans. We organize ourselves into various kinds of social groupings, such as nomadic bands, villages, cities, and countries, in which we work, trade, play, reproduce, and interact in many other ways. Unlike other species, we combine socialization with deliberate changes in social behavior and organization over time. Consequently, the patterns of human society differ from place to place and era to era and across cultures, making the social world a very complex and dynamic environment.

Insight into human behavior comes from many sources. The views presented here are based principally on scientific investigation, but it should also be recognized that literature, drama, history, philosophy, and other nonscientific disciplines contribute significantly to our understanding of ourselves. Social scientists study human behavior from a variety of cultural, political, economic, and psychological perspectives, using both qualitative and quantitative approaches. They look for consistent patterns of individual and social behavior and for scientific explanations of those patterns. In some cases, such patterns may seem obvi-ous once they are pointed out, although they may not have been part of how most people consciously thought about the world. In other cases, the patterns—as revealed by scientific investigation—may show people that their long-held beliefs about certain aspects of human behavior are incorrect.

This chapter covers recommendations about human society in terms of individual and group behavior, social organizations, and the processes of social change. It is based on a particular approach to the subject: the sketching of a comprehensible picture of the world that is consistent with the findings of the separate disciplines within the social sciences—such as anthropology, economics, political science, sociology, and psychology—but without attempting to describe the findings themselves or the underlying methodologies.

Chapter 8 / **The Designed World**

Agriculture
Materials and Manufacturing
Energy Sources and Use
Communication
Information Processing
Health Technology

The world we live in has been shaped in many important ways by human action. We have created technological options to prevent, eliminate, or lessen threats to life and the environment and to fulfill social needs. We have dammed rivers and cleared forests, made new materials and machines, covered vast areas with cities and highways, and decided—sometimes willy-nilly—the fate of many other living things.

In a sense, then, many parts of our world are designed, shaped, and controlled—largely through the use of technology—in light of what we take our interests to be. We have brought the earth to a point where our future well-being will depend heavily on how we develop and use and restrict technology. In turn, that will depend heavily on how well we understand the workings of technology and the social, cultural, economic, and ecological systems within which we live.

This chapter sets forth recommendations about certain key aspects of technology, with emphasis on the major human activities that have shaped our environment and lives. The chapter focuses on eight basic technology areas: agriculture, materials, manufacturing, energy sources, energy use, communication, information processing, and health technology.

Chapter 9 / **The Mathematical World**

Numbers
Symbolic Relationships
Shapes
Uncertainty
Reasoning

Mathematics is essentially a process of thinking that involves building and applying abstract, logically connected networks of ideas. These ideas often arise from the need to solve problems in science, technology, and everyday life—problems ranging from how to model certain aspects of a complex scientific problem to how to balance a checkbook.

This chapter presents recommendations about basic mathematical ideas, especially those with practical application, that together play a key role in almost all human endeavors. In Chapter 2, mathematics is characterized as a modeling process in which abstractions are made and manipulated and the implications are checked out against the original situation. Here, the focus is on five examples of the kinds of mathematical patterns that are available for such modeling: the nature and use of numbers, symbolic relationships, shapes, uncertainty (including probability, summarizing data, and sampling data), and reasoning.

Chapter 10 / **Historical Perspectives**

Displacing the Earth from the Center of the Universe
Uniting the Heavens and Earth
Relating Matter & Energy and Time & Space
Extending Time
Moving the Continents
Understanding Fire
Splitting the Atom
Explaining the Diversity of Life
Discovering Germs
Harnessing Power

There are two principal reasons for including some knowledge of history among the recommendations. One reason is that generalizations about how the scientific enterprise operates would be empty without concrete examples. Consider, for example, the proposition that new ideas are limited by the context in which they are conceived; are often rejected by the scientific establishment; sometimes spring from unexpected findings; and usually grow slowly, through contributions from many different investigators. Without historical examples, these generalizations would be no more than slogans, however well they might be remembered. For this purpose, any number of episodes might have been selected.

A second reason is that some episodes in the history of the scientific endeavor are of surpassing significance to our cultural heritage. Such episodes certainly include Galileo's role in changing our perception of our place in the universe; Newton's demonstration that the same laws apply to motion in the heavens and on earth; Darwin's long observations of the variety and relatedness of life forms that led to his postulating a mechanism for how they came about; Lyell's careful documentation of the unbelievable age of the earth; and Pasteur's identification of infectious disease with tiny organisms that could be seen only with a microscope. These stories stand among the milestones of the development of all thought in Western civilization.

All human cultures have included study of nature—the movement of heavenly bodies, the behavior of animals, the properties of materials, the medicinal properties of plants. The recommendations in this chapter focus on the development of science, mathematics, and technology in Western culture, but not on how that development drew on ideas from earlier Egyptian, Chinese, Greek, and Arabic cultures. The sciences accounted for in this report are largely part of a tradition of thought that happened to develop in Europe during the last 500 years—a tradition to which people from all cultures contribute today.

Chapter 11 / **Common Themes**

Systems
Models
Constancy and Change
Scale

Some important themes pervade science, mathematics, and technology and appear over and over again, whether we are looking at an ancient civilization, the human body, or a comet. They are ideas that transcend disciplinary boundaries and prove fruitful in explanation, in theory, in observation, and in design.

This chapter presents recommendations about some of those ideas and how they apply to science, mathematics, and technology. Here, thematic ideas are presented under four main headings: systems, models, constancy and change, and scale.

Chapter 12 / **Habits of Mind**

Values and Attitudes
Computation and Estimation
Manipulation and Observation
Communication
Critical-Response Skills

Throughout history, people have concerned themselves with the transmission of shared values, attitudes, and skills from one generation to the next. All three were taught long before formal schooling was invented. Even today, it is evident that family, religion, peers, books, news and entertainment media, and general life experiences are the chief influences in shaping people's views of knowledge, learning, and other aspects of life. Science, mathematics, and technology—in the context of schooling—can also play a key role in the process, for they are built upon a distinctive set of values, they reflect and respond to the values of society generally, and they are increasingly influential in shaping shared cultural values. Thus, to the degree that schooling concerns itself with values and attitudes—a matter of great sensitivity in a society that prizes cultural diversity and individuality and is wary of ideology—it must take scientific values and attitudes into account when preparing young people for life beyond school.

Similarly, there are certain thinking skills associated with science, mathematics, and technology that young people need to develop during their school years. These are mostly, but not exclusively, mathematical and logical skills that are essential tools for both formal and informal learning and for a lifetime of participation in society as a whole.

Taken together, these values, attitudes, and skills can be thought of as habits of mind because they all relate directly to a person's outlook on knowledge and learning and ways of thinking and acting.

This chapter presents recommendations about values, attitudes, and skills in the context of science education. The first part of the chapter focuses on four specific aspects of values and attitudes: the values inherent in science, mathematics, and technology; the social

value of science and technology; the reinforcement of general social values; and people's attitudes toward their own ability to understand science and mathematics. The second part of the chapter focuses on skills related to computation and estimation, to manipulation and observation, to communication, and to critical response to argument.

Chapter 13 / **Effective Learning and Teaching**
Principles of Learning
Teaching Science, Mathematics, and Technology
Although *Science for All Americans* emphasizes what students should learn, it also recognizes that how science is taught is equally important. In planning instruction, effective teachers draw on a growing body of research knowledge about the nature of learning and on craft knowledge about teaching that has stood the test of time. Typically, they consider the special characteristics of the material to be learned, the background of their students, and the conditions under which the teaching and learning are to take place.

This chapter presents—nonsystematically and with no claim of completeness—some principles of learning and teaching that characterize the approach of such teachers. Many of those principles apply to learning and teaching in general, but clearly some are especially important in science, mathematics, and technology education. For convenience, learning and teaching are presented here in separate sections, even though they are closely interrelated.

Chapter 14 / **Reforming Education**
The Need for Reform
Reform Premises
Project 2061 is concerned more with lasting reform of education than with the immediate improvement of the schools—although such improvement is certainly needed, possible, and under way in many parts of the United States. But, as the nation discovered after Sputnik more than three decades ago, enduring educational reform is not easily achieved.

The possibility of successfully restructuring science education in its entirety depends on the presence of a public demand for reform in science education and on what we as a nation think it takes to achieve reform. This chapter begins by showing that there is in fact a consensus on the need for reform in science, mathematics, and technology education, and then presents the premises that underlie the approach of Project 2061 to reform.

Chapter 15 / **Next Steps**
Project 2061
An Agenda for Action
The Future
Science for All Americans has little to say about what ails the educational system, points no finger of blame, prescribes no specific remedies. Rather, it attempts to contribute substantially to educational reform by serving as a starting point for two sets of critical, reform-oriented actions.

One set is based on use of the report as the first step in a multistage, long-term developmental process. *Science for All Americans* should be used as the conceptual basis for recommendations for change in all parts of the educational system.

The other set of actions is based on the fact that the report provides a new and unusually substantive opportunity for everyone who has a stake in educational reform to reappraise the progress made so far, redirect their efforts as needed, and recommit themselves to fundamental reform goals.

This final chapter of *Science for All Americans* starts with a brief outline of the next steps toward reform being taken by Project 2061. It then explores some of the ways in which the report can be put to work by educators, policymakers, and the interested public.

WAYNE THIEBAUD, *Nine Books, 1991-92.*

Chapter **2** SCIENCE TRADE BOOKS

Teachers understand the importance of lifelong learning. And teachers of science, mathematics, and technology may be more motivated than most to extend their knowledge because of the large territory (and continuing expansion) of those fields. In addition to formal coursework, there is a wide variety of books, usually written for a general audience, that delve fairly deeply into particularly interesting scientific topics. Many of these books are written by well-known scientists and engineers who are also accomplished writers, able to make complex ideas and technologies comprehensible to the lay reader.

The _Professional Development_ CD-ROM provides more than 200 citations identifying a variety of "trade books" (as distinct from textbooks or reference books), that are likely to enrich the reader's understanding of science literacy goals as described in _Science for All Americans (SFAA)_. Teachers can follow their own interests in browsing through as few or as many as they wish, or they can search more systematically for books dealing with a specific idea or topic.

Most teachers will be quite knowledgeable about some of the topics in _SFAA_ but less familiar with others. Therefore, the trade books selected range from basic introductions—some actually written with young adults in mind—to more sophisticated treatments requiring considerable background knowledge. In addition to nonfiction science books, the database includes novels, philosophical works, and collections of essays that shed light on many aspects of science literacy.

The database does not include general reference books that provide only a brief overview of topics from many chapters of _SFAA_, although there are some excellent books of that genre[1]. Treatment of topics in this kind of book is inevitably condensed and may not be adequate to improve understanding. (Indeed, _SFAA_ itself is too concise to be useful for actual learning of the ideas described in it.) With only a few exceptions, the database does not include textbooks, as they typically have far more technical language and content than _SFAA_ recommends as essential to science literacy.

---

[1]If you would like a more general introduction to the broad range of topics covered in _Science for All Americans_, these three reference works may be useful: _Science Matters: Achieving Scientific Literacy_ by Robert M. Hazen and James Trefil, (Doubleday, 1991); _How the World Works: A Guide to Science's Greatest Discoveries_ by Boyce Rensberger (Morrow, 1986); and _Young Person's Guide to Science: Ideas That Change the World_ by Roy A. Gallant (Macmillan, 1993).

To identify most of the trade books for the database, Project 2061 staff consulted *Science Books & Films*, a periodical published nine times a year by the American Association for the Advancement of Science to provide critical reviews of resources intended for use in science, mathematics, and technology education. Project 2061 also consulted relevant bibliographies of recommended reading, such as "The Literature of Science" by Gregg Sapp in the March 1995 issue of *The Library Journal*. Professionals in the fields of science, mathematics, and technology reviewed the list of prospective database entries, suggesting some additional trade books to include.

SELECTION CRITERIA

Every trade book selected met the following criteria:

- It explicitly addressed content in *SFAA*, although it might include other material as well.
- It was highly rated (one or two stars) by *Science Books & Films* or a similar source.
- *Science Books & Films* or Project 2061's own reviewers indicated that it would be interesting and informative to a general audience, including teachers of all grades and all subjects.

The trade-book database identifies a sampling of the many science trade books that have been published over the past decade, along with a few older books that are now generally considered classics. A special attempt was made to identify trade books for *SFAA* topics that are not standard classroom fare—for example, technology and the history of science. Although several of the recommended trade books are out of print, they should be available through local public or university libraries.

NOTES ON THE TRADE-BOOK DATABASE ON CD-ROM

Searching. In using the trade-book database, you have the option of browsing the entire collection of entries alphabetically by author or title, or calling up books related to a particular chapter or section of *SFAA*. The information icon available on the main trade-book screen leads to brief introductions to the trade books selected for each chapter of *SFAA*, followed by notes on special characteristics of some of the books. Many of the trade books relate to more than one chapter or section of *SFAA*, and so appear on more than one list.

Links to *Science for All Americans*. The database identifies the *most significant* links between the trade books' content and specific *SFAA* chapters and/or sections, but there are other connections that may be rewarding to explore. A trade book linked to a particular section may address only some of the topics in that section and may also address topics not included in *SFAA* at all.

Reviews. Whenever *Science Books & Films* reviewed a book, that review was included in the database. In other cases, we included a review from a different source. The reviews from *Science Books & Films* identify appropriate audiences by the codes GA, JH, YA, C, T, and P; these indicate General Audience, Junior High School Students, Young Adults, College Students, Teachers, and Professionals, respectively. (Reviews published in *Science Books & Films* prior to the September/October 1986 issue used the code SH to identify books aimed at high school students.) Most of the books in the database target General and Teacher audiences.

Additional Reading. The trade-book database as a whole is a good starting point for exploration of science literacy. To develop a more sophisticated understanding of a particular idea found in *SFAA*, users may find it beneficial to discuss the idea with colleagues and read about it in several different contexts. To pursue ideas of special interest at a more advanced level, the bibliographies contained in the books themselves can be helpful for locating additional books and articles.

Examples from the CD-ROM. All of the trade books included on the *Professional Development* CD-ROM are listed on the left-hand pages that follow and are organized under the twelve *SFAA* chapters to which they are linked. A sample book entry selected for each chapter is presented on right-hand pages and includes the book's cover and table of contents, links to relevant chapters and sections of *SFAA* (e.g., 1A indicates a link to *SFAA* Chapter 1: The Nature of Science, section A, the Scientific World View*),* one complete or excerpted review of the book, and full bibliographic information.

For a description of the topics covered in each chapter of *SFAA*, see this volume's Chapter 1: About *Science for All Americans*, beginning on page 3.

ALSO SEE

Science for All Americans / Chapter 1: The Nature of Science

Animal Experimentation: Cruelty or Science?	Nancy Day	Enslow	1994
Apprentice to Genius: The Making of a Scientific Dynasty	Robert Kanigel	Macmillan	1986
Bad Science	Gary Taubes	Random House	1993
Bitten by the Biology Bug	Maura C. Flannery	National Association of Biology Teachers	1991
Blueprints: Solving the Mystery of Evolution	Maitland A. Edey and Donald C. Johanson	Little, Brown	1989
The Common Sense of Science	Jacob Bronowski	Harvard University Press	1978
Great Essays in Science	Martin Gardner (Ed.)	Prometheus Books	1994
How We Know: An Exploration of the Scientific Process	Martin Goldstein and Inge F. Goldstein	Plenum	1978
I Want to be a Mathematician: An Automathography	Paul R. Halmos	Springer-Verlag	1985
Late Night Thoughts on Listening to Mahler's Ninth Symphony	Lewis Thomas	Viking	1983
Mayonnaise and the Origin of Life: Thoughts of Minds and Molecules	Harold J. Morowitz	Charles Scribner's Sons	1985
A Physicist on Madison Avenue	Tony Rothman	Princeton University Press	1991
Radioactivity: From the Curies to the Atomic Age	Tom McGowen	Franklin Watts Inc.	1986
Rats, Lice and History	Hans Zinsser	Little, Brown	1934
The Science Gap: Dispelling the Myths and Understanding the Reality of Science	Milton A. Rothman	Prometheus Books	1992
The Scientific Attitude, 2nd ed.	Frederick Grinnell	The Guilford Press	1992
The Search for Solutions	Horace Freeland Judson	Holt, Rinehart and Winston	1980
Signs of Life	Robert Pollack	Houghton Mifflin	1994
To Know a Fly	Vincent Gaston Dethier	Holden-Day, Inc.	1963
The Virgin and the Mousetrap: Essays in Search of the Soul of Science	Chet Raymo	Viking	1991
The World of Mathematics	James R. Newman	Simon & Schuster	1956

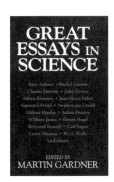

SFAA link: 1A 1B 1C

Gardner, Martin (Ed.)

Great Essays in Science

Prometheus Books

1994

427 pp.

$16.95 (paper)

0-87975-853-8

C, GA **

"How do you condense, into 427 pages, some of the best writings in science during the last 100 years? Select the 31 most wonderful contributors to science and science writing in that period. Pick some of the most thought-provoking contributions by them that represent the peak of their accomplishments. That is what Martin Gardner has done in editing this anthology of *Great Essays in Science*. Although, as a matter of fact, these are not all essays in science, they are all about science, society, and nature. Two pieces are from works of fiction, but they keep their scientific character and thrust all the same. On the old (but not continuing) debate on the relative importance of the roles of literature versus science, there are four essays. Jose Ortega Y Gasset and John Burroughs speak for the lesser role for science, while Thomas Huxley and Isaac Asimov present even more forceful arguments for the other side. From the world of insects (Jean Henri Fabre and Maurice Maeterlinck) to the organisms under the deep oceans beyond the reach of sunlight, the essays cover some of the most fascinating topics that have ever evoked the curiosity of those who have experienced the beauty of science. The essays by Stephen Jay Gould (Nonmoral Nature) and Carl Sagan (Reflections on a Grain of Salt) show that the intellectual giants of the 20th century can match and surpass their counterparts of earlier centuries in the vastness of the sweep of their imagination and intellectual excellence. If you have thought about what the seven wonders of the world in the 20th century would be, who else would be better suited to offer the best answer but Lewis Thomas? In his typically incisive and provocative fashion, he offers the following list: the bacteria that thrive under extreme pressures and temperatures in deep sea vents, the beetle *oncideres*, the scrapie virus—the strangest thing in all biology, the olfactory receptor cells of animals, the termites as a society of insects, the human child, and the planet earth. If you are curious as to what his arguments for selecting just these items are, you will want to read this book."
—Reviewed by B. Thyagarajan in *Science Books & Films*, 30/3 (April 1994), p. 76.

With additional essays by

Francis Bacon	Jonathan Norton Leonard	Laura Fermi
John Dewey	J. Robert Oppenheimer	Samuel Goudsmit
William James	Alfred North Whitehead	Robert Louis Stevenson
Havelock Ellis	John Dos Passos	Sigmund Freud
Gilbert Keith Chesterton	Julian Huxley	Bertrand Russell
Joseph Wood Krutch	Arthur Stanley Eddington	Albert Einstein
Ernest Nagel	Aldous Huxley	
	Rachel Carson	
	H.G. Wells	

Science for All Americans / CHAPTER 2: THE NATURE OF MATHEMATICS

Archimedes' Revenge: The Joys and Perils of Mathematics	Paul Hoffman	Fawcett Crest	1988
How We Know: An Exploration of the Scientific Process	Martin Goldstein and Inge F. Goldstein	Plenum	1978
I Want to Be a Mathematician: An Automathography	Paul R. Halmos	Springer-Verlag	1985
The Mathematical Universe	William Dunham	John Wiley & Sons	1994
One Two Three...Infinity	George Gamow	Dover	1947
Recent Revolutions in Mathematics	Albert Stwertka	Franklin Watts Inc.	1987
The Refrigerator and the Universe: Understanding the Laws of Energy	Martin Goldstein and Inge F. Goldstein	Harvard University Press	1993
The World of Mathematics	James R. Newman	Simon & Schuster	1956

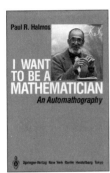

SFAA link: 1C 2A 2C

Halmos, Paul R.
*I Want to Be a
Mathematician: An
Automathography*
(Illus.)
Springer-Verlag
1985
410 pp.
$52.00
0-387-96078-3
Index

"At last we have a thorough account (one that stands the test of re-reading, and the only such, in fact) of the period that runs approximately from the forties to the present-day, a period that may go down in history as one of the golden ages of mathematics. However, the theme that emerges from this collection of amusing anecdotes is not the welcome lesson we would expect as the bequest of a golden age. Halmos's tales of incompetent department heads, of Neanderthal deans, of obnoxious graduate students unwittingly reveal, in the glaring light of gossip, the constant bungling, the lack of common sense, the absence of savoir faire that is endemic in mathematics departments everywhere. Take, for example, the turning point of the author's career, the incident of his leaving the University of Chicago. Even granting Halmos's contention that his papers may have lacked depth (at least, in someone's opinion) in comparison with those of certain colleagues of his (a debatable thesis, then and now), it still seems clear that the university made a mistake by dispensing with Halmos's services. Whatever his other merits, Halmos is now regarded as the best expositor of mathematics of his time. His textbooks have had an immense influence on the development of mathematics since the fifties, especially by their influence on mathematicians in their formative years. Halmos's glamor would have been a far sounder asset to the University of Chicago than the deep but dull results of an array of skillful artisans. What triumphed at the time is an idea that still holds sway in mathematics departments today, namely, the simplistic view of mathematics as a linear progression of problems solved and theorems proved, in which any other function that may contribute to the well-being of the field (most significantly, that of exposition) is to be valued roughly on a par with that of a janitor. It is as if in the filming of a movie all credits were to be granted to the scriptwriter, at the expense of other contributors (actors, directors, costume designers, musicians, etc.) whose roles are equally essential for the movie's success."
—Reviewed by Gian-Carlo Rota in *The American Mathematical Monthly*, 94 (August/September 1987), p. 700.

Contents

Part 1: Student

Reading and Writing
   and 'Rithmetic
A College Education
Graduate School
Learning to Study
Learning to Think
The Institute
Winning the War

Part II: Scholar

A Great University
The Early Years
The Fabulous Fifties

Part III: Senior

How to Teach
To Sydney, to Moscow, and Back
How to Do Almost Everything
Service, One Way or Another
Coda

Science for All Americans / Chapter 3: The Nature of Technology

The Control Revolution: Technical and Economic Origins of the Information Society	James R. Beniger	Harvard University Press	1986
The Day the Sun Rose Twice	Ferenc Morton Szasz	University of New Mexico Press	1984
The Design of Everyday Things	Donald A. Norman	Doubleday (originally *The Psychology of Everyday Things*, Basic Books)	1988
Discovery, Innovation and Risk: Case Studies in Science and Technology	Newton H. Copp and Andrew W. Zanella	MIT Press	1993
Engineering and the Mind's Eye	Eugene S. Ferguson	MIT Press	1992
Engines of Change: The American Industrial Revolution, 1790–1860	Brooke Hindle and Steven Lubar	Smithsonian Institution Press	1986
Flying Buttresses, Entropy, and O-Rings: The World of an Engineer	James L. Adams	Harvard University Press	1992
Inventors at Work: Interviews with 16 Notable American Inventors	Kenneth A. Brown	Tempus	1988
Medical Technology and Society	Joseph D. Bronzino, Vincent H. Smith, and Maurice L. Wade	McGraw-Hill	1990
The Nuclear Energy Option: An Alternative for the '90s	Bernard L. Cohen	Plenum	1990
The Search for Solutions	Horace Freeland Judson	Holt, Rinehart and Winston	1980
Supercomputing and the Transformation of Science	William J. Kaufmann III and Larry L. Smarr	*Scientific American* Library	1993
Superstuff!	Fred Bortz	Franklin Watts Inc.	1990
Technologies without Boundaries: On Telecommunications in a Global Age	Ithiel De Sola Pool	Harvard University Press	1990
Technology and the Future, 6th ed.	Albert H. Teich (Ed.)	St. Martin's Press	1993
To Engineer Is Human	Henry Petroski	Random House	1982
Works of Man	Ronald W. Clark	Viking	1985

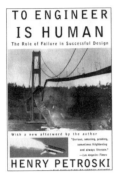

TO ENGINEER IS HUMAN

The Role of Failure in Successful Design

With a new afterword by the author

"Serious, amazing, probing, sometimes frightening and always literate." —Los Angeles Times

HENRY PETROSKI

SFAA link: 3B

Petroski, Henry

To Engineer is Human

(Illus.)

Random House

1982

251 pp.

$11.00

0-679-73416-3

Bibliography; Index

"Here is a gem of a book. Engineering professor Petroski raises the concept that past failure in engineering design is the handmaiden of future success and innovation. He discusses some monumental failures—like the collapse of elevated walkways in a Kansas City hotel—and shows how they led engineers to advance their art to meet new needs. One chapter declares, 'Falling Down Is Part of Growing Up'. His examples are mostly the honest-mistake kind, and not the sloppy design and testing, for instance, that results in recalls of new autos. But in marvelously clear prose, he gives valuable insight into the limits of engineering and its practitioners. A fine book for general and history-of-technology collections alike."
—Reviewed by Daniel LaRossa in *Library Journal*, 110 (September 1, 1985), p. 206.

Contents

Being Human

Falling Down Is Part of Growing Up

Lessons from Play; Lessons from Life

Engineering As Hypothesis

Success Is Foreseeing Failure

Design Is Getting from Here to There

Design as Revision

Accidents Waiting to Happen

Safety in Numbers

When Cracks Become Breakthroughs

Of Bus Frames and Knife Blades

Interlude: The Success Story of the Crystal Palace

The Ups and Downs of Bridges

Forensic Engineering and Engineering Fiction

From Slide Rule to Computer: Forgetting How
     It Used to Be Done

Connoisseurs of Chaos

The Limits of Design

Appendix: "The Deacon's Masterpiece"
     by Oliver Wendell Holmes

Science for All Americans / CHAPTER 4: THE PHYSICAL SETTING

Atom: Journey Across the Subatomic Cosmos	Isaac Asimov	Dutton/Signet	1991
Beginnings: The Story of Origins—Of Mankind, Life, the Earth, the Universe	Isaac Asimov	Walker & Co.	1987
An Equation That Changed the World: Newton, Einstein, and the Theory of Relativity	Harald Fritzsch and translated by Karin Heusch	University of Chicago Press	1994
From Stone to Star: A View of Modern Geology	Claude Allegre and translated by Deborah Kurmes Van Dam	Harvard University Press	1992
Isaac Asimov's Guide to Earth and Space	Isaac Asimov	Random House	1991
Knowledge and Wonder—The Natural World as Man Knows It	Victor F. Weisskopf	Doubleday	1979
Lonely Hearts of the Cosmos: The Story of the Scientific Quest for the Secret of the Universe	Dennis Overbye	HarperCollins	1991
Nuclear Choices: A Citizen's Guide to Nuclear Technology	Richard Wolfson	MIT Press	1991
A Physicist on Madison Avenue	Tony Rothman	Princeton University Press	1991
Physics for Poets	Robert H. March	McGraw-Hill	1992
Physics: From Newton to the Big Bang	Albert and Eve Stwertka	Franklin Watts Inc.	1986
Planet Earth	Jonathan Weiner	Bantam Books	1986
Powers of Ten: A Book about the Relative Size of Things in the Universe and the Effect of Adding Another Zero	Philip and Phylis Morrison and the office of Charles and Ray Eames	*Scientific American* Library	1994
Recent Revolutions in Mathematics	Albert Stwertka	Franklin Watts Inc.	1987
The Refrigerator and the Universe: Understanding the Laws of Energy	Martin Goldstein and Inge F. Goldstein	Harvard University Press	1993
Seven Ideas That Shook the Universe	Nathan Spielberg and Bryon D. Anderson	John Wiley & Sons	1987
Stephen Hawking's Universe	John Boslough	Quill/Morrow	1985
Superstuff!	Fred Bortz	Franklin Watts Inc.	1990

SFAA link: 1A 1B 1C
4A 4B 4C 4D 4E
4F 4G 12A
Rothman, Tony
A Physicist on Madison Avenue
(Illus.)
Princeton University Press
1991
141 pp.
$19.95
0-691-08731-8
Index
T, GA **

"The somewhat unusual title of this book is intended to indicate that the author, a physicist, has temporarily assumed the guise of a science journalist. The book consists of nine chapters, each an article on science written for the educated layperson. Six of the articles were published previously in the popular science magazines *Discover* and *Scientific American*. In the first chapter, the reader is introduced to a physicist's method of treating data. The author recounts how his recognition of the similarity of a histogram of data on magazine sales to a Gaussian distribution enabled him to conclude that newsstand sales were dependent on random effects.... Chapter 2 continues the discussion of the great value of the scientific approach by describing how the difficult problem of improving the quality of musical instruments may be attacked by using physical theory to guide experimentation. The next chapter is concerned with the fact that no arrow of time exists in physical theories, and yet the forward march of time is evident in almost everything that we do.... Chapter 4 discusses the role of life in the evolution of the universe since the Big Bang. The basic issue is the extent to which the constants of nature must be compatible with the existence of carbon-based Homo sapiens. The next three articles concern the Big Bang theory of cosmology. The beginning of the universe, our present situation, and our future are considered. A penetrating critique of the Big Bang theory is given, and alternative cosmologies are proposed. Chapter 8 describes a puzzling object in the sky called Cygnus X3. Its emissions cannot be explained by present theory. The last article appeared as an April Fool's joke in *Scientific American*. Nevertheless, it contains an excellent review of natural constraints on the construction of particle accelerators, such as the Superconducting Super Collider.... This book has something for everyone. It is well written, authoritative, interesting, and informative. It contains much good physics without relying on mathematics...an excellent introduction to modern cosmology for a general audience."
—Reviewed by Reuben Benumof in *Science Books & Films*, 27/7 (October 1991), p. 200.

Contents

A Physicist on Madison Avenue

Instruments of the Future, Traditions of the Past

The Seven Arrows of Time

The Measure of All Things

On That Day, When the Earth Is Dissolved
   in Positrons...

The Epoch of Observational Cosmology

Alternative Cosmologies
   (with G.F.R. Ellis and Richard Matzner)

Stranger Than Fiction: Cygnus X-3
   (with David Aschman)

The Ultimate Collider Antoni Akahito
   (Tony Rothman)

Science for All Americans / Chapter 5: The Living Environment

The Beak of the Finch: A Story of Evolution in Our Time	Jonathan Weiner	Knopf	1994
Beauty and the Beast: The Coevolution of Plants and Animals	Susan Grant	Charles Scribner's Sons	1984
Beginnings: The Story of Origins—Of Mankind, Life, the Earth, the Universe	Isaac Asimov	Walker & Co.	1987
Bitten by the Biology Bug	Maura C. Flannery	National Association of Biology Teachers	1991
Blueprints: Solving the Mystery of Evolution	Maitland A. Edey and Donald C. Johanson	Little, Brown	1989
The Body	Anthony Smith	Viking	1986
Cells	George S. Fichter	Franklin Watts Inc.	1986
The Creation of Life: Past, Future, Alien	Andrew Scott	Basil Blackwell	1986
Diatoms to Dinosaurs: The Size and Scale of Living Things	Chris McGowan	Island Press	1994
Ever Since Darwin: Reflections in Natural History	Stephen Jay Gould	W. W. Norton	1977
Extinction	Rebecca Stefoff	Chelsea House Publishers	1992
The Flamingo's Smile: Reflections in Natural History	Stephen Jay Gould	W. W. Norton	1985
The Flight of the Iguana: A Sidelong View of Science and Nature	David Quammen	Delacorte Press	1988
Gene Future: The Promise and Perils of the New Biology	Thomas F. Lee	Plenum	1993
Global Ecology	Colin Tudge	Oxford University Press	1991
Knowledge and Wonder—The Natural World as Man Knows It	Victor F. Weisskopf	Doubleday	1979
The Language of Genes	Steve Jones	Anchor/Doubleday	1994
The Lives of a Cell: Notes of a Biology Watcher	Lewis Thomas	Viking	1974
Mammal Evolution: An Illustrated Guide	R. J. G. Savage	Facts On File	1986
Mayonnaise and the Origin of Life: Thoughts of Minds and Molecules	Harold J. Morowitz	Charles Scribner's Sons	1985
The New Biology: Discovering the Wisdom in Nature	Robert Augros and George Stanciu	Shambhala	1987
New Theories on the Origins of the Human Race	Christopher Lampton	Franklin Watts Inc.	1989
Signs of Life	Robert Pollack	Houghton Mifflin/Viking	1994
A View from the Heart: Bayou Country Ecology	June C. Kennedy	Blue Heron Press	1991
The Virgin and the Mousetrap: Essays in Search of the Soul of Science	Chet Raymo	Viking	1991
Was George Washington Really the Father of Our Country? A Clinical Geneticist Looks at World History	Robert Marion	Addison-Wesley	1994
What Makes You What You Are: A First Look at Genetics	Sandy Bornstein	Messner	1989
The World of Microbes	Howard Gest	Science Tech	1987

SFAA link: 5F 10H
Weiner, Jonathan
*The Beak of the Finch:
A Story of Evolution
in Our Time*
(Illus.)
Alfred A. Knopf Inc.
1994
332 pp.
$25.00
0-679-40003-6
Index
C, T, GA**

"…This book clearly and elegantly shows the modern-day reality of the visible action of evolution as observable fact. The text is based on numerous scientific publications by professors Peter and Rosemary Grant and their associates. Weiner takes us through the steps of these two researchers in over a decade of daily observations and measurements on perhaps the most famous birds in the world, Darwin's finches. Darwin himself was the first to describe and marvel over this diverse, yet similar, group of closely related birds that inhabit the Galapagos Islands in a patchwork of species and adaptations. Darwin was also the first to note how seemingly minute variations in the beaks of different species coincided with very different behaviors and distributions on the islands. The Grants have taken Darwin's and others' observations many steps further…. What they found first was that the population of various finches on Daphne Major are not static. The seasonal differences swing wildly about and are directly related to changes in rainfall, food availability, and other factors. Detailed studies show that seemingly minute differences of only a single millimeter in the depth of a finch's beak can strongly influence that bird's survival, breeding potential, and contribution to the gene pool. The researchers were able to observe conditions that favored both greater separation of species (i.e., greater distinction) and the fusion of species (when conditions favor the survival of hybrids). And, unlike Darwin's wildest dreams, the Grants actually measured factors that might easily suggest the origin of new species. Their students who were working with blood samples were able to pinpoint DNA chains that corresponded with changes in the birds' structure and changes in climate…. The author relates all of these factors to the readers through various other examples of plants, animals, and humans in a highly readable manner."
—Reviewed by James W. Waddick in *Science Books & Films*, 30/9 (December 1994), pp. 263-4.

Contents

Science for All Americans / CHAPTER 6: THE HUMAN ORGANISM

The Ascent of Man	Jacob Bronowski	Little, Brown	1976
Beginnings: The Story of Origins—Of Mankind, Life, the Earth, the Universe	Isaac Asimov	Walker & Co.	1987
Bitten by the Biology Bug	Maura C. Flannery	National Association of Biology Teachers	1991
Blueprints: Solving the Mystery of Evolution	Maitland A. Edey and Donald C. Johanson	Little, Brown	1989
The Body	Anthony Smith	Viking	1986
The Body in Time	Kenneth Jon Rose	John Wiley & Sons	1988
The Brain	Richard M. Restak	Bantam Books	1984
Ever Since Darwin: Reflections in Natural History	Stephen Jay Gould	W. W. Norton	1977
Global Ecology	Colin Tudge	Oxford University Press	1991
Late Night Thoughts on Listening to Mahler's Ninth Symphony	Lewis Thomas	Viking	1983
The Lives of a Cell: Notes of a Biology Watcher	Lewis Thomas	Viking	1974
Mammal Evolution: An Illustrated Guide	R. J. G. Savage	Facts On File	1986
The Man Who Mistook His Wife for a Hat	Oliver Sacks	Harper & Row	1985
Microbe Hunters	Paul de Kruif	Harcourt Brace Jovanovich	1926
The Mind	Anthony Smith	Viking	1984
The New Biology: Discovering the Wisdom in Nature	Robert Augros and George Stanciu	Shambhala	1987
New Theories on the Origins of the Human Race	Christopher Lampton	Franklin Watts Inc.	1989
Rats, Lice and History	Hans Zinsser	Little, Brown	1934
Traces of Life: The Origins of Humankind	Kathryn Lasky	William Morrow & Co.	1989
The Virus Invaders	Alan E. Nourse	Franklin Watts Inc.	1992
Was George Washington Really the Father of Our Country? A Clinical Geneticist Looks at World History	Robert Marion	Addison-Wesley	1994
What Makes You What You Are: A First Look at Genetics	Sandy Bornstein	Messner	1989
The World of Microbes	Howard Gest	Science Tech	1987
The Youngest Science: Notes of a Medicine-Watcher	Lewis Thomas	Viking	1983

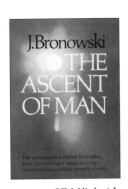

SFAA link: 6A
Bronowski, Jacob
The Ascent of Man
(Illus.)
Little, Brown
1976
448 pp.
$29.95
0-316-10933-9
Bibliography; Index

"Bronowski examines the coincidence of nature, culture, and a scientist which brought about what he considers the most significant inventions—the change from nomadic to agricultural life, the use of metals, calculus, architecture, astronomy, power capturing machines, the theories of natural selection, genes, atoms, nuclear physics, and relativity....[This] is an inspiring account of man in his prouder moments being true to himself. It provides the reader with a viewpoint by which he can make some sense out of the present knowledge explosion. It organizes the bits of half-thoughts with which we are bombarded daily and provides us with a challenge of our own if we are to be involved in the ascent of man."
—Reviewed by Nina Franco in *Best Seller*, 34/115 (June 1, 1974). (From *Book Review Digest*, 1974).

"This book could have been written only by Jacob Bronowski, for few have a vision that is at once extensive and intensive and that brings together the talents of mathematician, statistician, historian, teacher, inventor, and poet. Bronowski recites in a fluent and pleasant style the scientific and intellectual history of Man—a history that reveals Man as unique among the animal species. This is an informal account of Man's search to understand Nature, and it is the human qualities of thought and imagination that have led Man to analyze the physical world...."
—*Choice*, 11:782 (July/August 1974). (From *Book Review Digest*, 1974).

"Dr. Bronowski's view of science history is highly personal. To glorify it as he does may seem naively optimistic to those who wonder if man will smother himself in overpopulation or exhaust his resources in undisciplined growth. But I share [his] faith that the unfolding understanding he chronicles represents a spiritual growth that can liberate mankind from such threats. The book reflects other aspects of its author besides this underlying conviction. There is Bronowski the poet...the champion of integrity [and]...the science chauvinist, who can unselfconsciously declare: '...the intellectual leadership of the 20th century rests with scientists.' Since Lord Clark might well claim equivalent status for artists, we can forgive Dr. Bronowski his bias."
—Reviewed by R. C. Cowen in *Christian Science Monitor*, (June 25, 1974). (From *Book Review Digest*, 1974).

Contents

Science for All Americans / CHAPTER 7: HUMAN SOCIETY

The Control Revolution: Technical and Economic Origins of the Information Society	James R. Beniger	Harvard University Press	1986
Economics Explained: Everything You Need to Know about How the Economy Works and Where It's Going	Robert L Heilbroner and Lester C. Thurow	Simon & Schuster/Touchstone	1994
Economy and Society	Robert J. Holton	Routledge	1992
A History of Private Life: Riddles of Identity in Modern Times	Antoine Prost and Gérard Vincent (editors), translated by Arthur Goldhammer	Harvard University Press	1991
The Human Cycle	Colin M. Turnbull	Simon & Schuster	1983
Man on Earth	John Reader	University of Texas Press	1988
Metaman: The Merging of Humans and Machines into a Global Superorganism	Gregory Stock	Simon & Schuster	1993
So Shall You Reap: Farming and Crops in Human Affairs	Otto T. Solbrig and Dorothy J. Solbrig	Island Press	1994

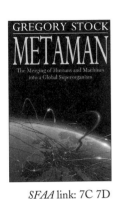

SFAA link: 7C 7D
7E 7G 11A
Stock, Gregory
*Metaman: The Merging of
Humans and Machines into
a Global Superorganism*
(Illus.)
Simon & Schuster
1993
320 pp.
$24.00
0-671-70723-X
Glossary; Index
C, T, GA **

"In this book, we are asked to look down from the moon at the night side of the earth and view 'the thin planetary patina of humanity and its creations' as a living entity, a superorganism consisting of interdependent human and nonhuman elements. This superorganism has the fundamentals of life: It has specialization and integration of the various parts, it metabolizes and converts energy into activity, it has an integrated response to the environment, and it evolves, albeit by reshaping itself rather than reproducing.... The book is written in an intelligent and entertaining manner, with numerous predictions of the future based on the interrelations among people and machines.... Examples of future technology, of which we presently see only hints, include genetically engineered plant viruses, massively parallel computing, and cochlear and other implants of machines into humans. This book is a fun read. There is optimism throughout. The author states that there is no further threat of global war because of our increasing interdependencies. It is also encouraging to note that we needn't transform human nature; the superorganism is able to accommodate various human deficiencies, since it is an overall entity.... It is not necessary to believe all the predictions in order to enjoy reading them.... Examples are given of unrealistic policies that we currently advocate, such as trying to achieve totally risk-free environments when, in so doing, we introduce other, perhaps worse, risks. Another example, which some might find offensive, is the policy of paying large amounts of money to treat very low-birthweight babies in intensive care when such care 'is based on unwarranted assumptions' and the money might be better spent in other ways."
—Reviewed by Sonja Johansen in *Science Books & Films*, 30/2 (March 1994), p. 40.

Contents

Planetary Superorganism:
Glimpses of a Promising Future
The Birth of Metaman:
Extending Life's Patterns
**A Fusion of Technology
and Biology:**
The Nonhuman Constituents of
Metaman
Inner Workings Explored:
Physiology of the Superorganism

The Mind of Metaman:
An Evolving Global Brain
The Foundation for Our Future:
Accelerating Toward
Global Union
Rites of Passage:
Perspectives on Global Concerns
Man and Metaman:
The Individual and the
Superorganism
Beyond Homo sapiens:
Intimations of Humanity's Future
Environmental Transformations:
The Changing Nature of Nature

The Trajectory of Social Change:
A Story of Rising Expectations
Power and Choice:
Challenges to Human Values
Darwin Extended:
The Evolution of Evolution
A New Mythology:
Metaman and Our Quest
for Meaning

Science for All Americans / CHAPTER 8: THE DESIGNED WORLD

The Age of Electronic Messages	John G. Truxal	MIT Press	1990
The Control Revolution: Technical and Economic Origins of the Information Society	James R. Beniger	Harvard University Press	1986
The Day the Sun Rose Twice	Ferenc Morton Szasz	University of New Mexico Press	1984
Discovery, Innovation and Risk: Case Studies in Science and Technology	Newton H. Copp and Andrew W. Zanella	MIT Press	1993
Energy Demands	Brian Gardiner	Franklin Watts Inc.	1990
Engines of Change: The American Industrial Revolution, 1790-1860	Brooke Hindle and Steven Lubar	Smithsonian Institution Press	1986
Medical Technology and Society	Joseph D. Bronzino, Vincent H. Smith, and Maurice L. Wade	McGraw-Hill	1990
Microbe Hunters	Paul de Kruif	Harcourt Brace Jovanovich	1926
Nuclear Choices: A Citizen's Guide to Nuclear Technology	Richard Wolfson	MIT Press	1991
The Nuclear Energy Option: An Alternative for the '90s	Bernard L. Cohen	Plenum	1990
Our Natural Resources and Their Conservation, 7th ed.	Harry Kircher, Donald Wallace, and Dorothy Gore	Interstate Publishers	1991
Pasteur and Modern Science	Rene Dubos	Science Tech	1988
So Shall You Reap: Farming and Crops in Human Affairs	Otto T. Solbrig and Dorothy J. Solbrig	Island Press	1994
Supercomputing and the Transformation of Science	William J. Kaufmann III and Larry L. Smarr	*Scientific American* Library	1993
Superstuff!	Fred Bortz	Franklin Watts Inc.	1990
Technologies without Boundaries: On Telecommunications in a Global Age	Ithiel De Sola Pool	Harvard University Press	1990
Technology and the Future, 6th ed.	Albert H. Teich (Ed.)	St. Martin's Press	1993
Telecommunications: From Telegraphs to Modems	Christopher Lampton	Franklin Watts Inc.	1991
The Virus Invaders	Alan E. Nourse	Franklin Watts Inc.	1992
Works of Man	Ronald W. Clark	Viking	1985

SFAA link: 7C 8A
Solbrig, Otto T., and
Dorothy J. Solbrig
So Shall You Reap:
Farming and Crops
in Human Affairs
(Illus.)
Island Press
1994
284 pp.
$27.50
1-55963-308-5
Index
C, T, GA *

"After describing in the first chapter the ways that early humans acquired food, the authors in the next chapter discuss methods by which hunter-gatherers evolved into farmers. Adaptation of rudimentary agricultural practices during prehistoric times are described in the third chapter. Domestication of plants for food and adoption of improved agricultural practices form part of the fourth and fifth chapters respectively. The sixth chapter is devoted to the colonization of Crete by people from Asia Minor and the spread of agricultural practices to Europe, where crop rotation and animal power were adopted. The seventh chapter explains various factors leading to the establishment of medieval farms, which changed agriculture from a subsistence to a market system. Cultivation of sugar cane in Atlantic islands, Brazil, and the Caribbean transformed farming for food into farming for trade and capital accumulation, as detailed in the eighth chapter. The ninth chapter explains ways that farming practices improved in the Americas, which led to the exchange of produce between the Old and New Worlds. In the tenth chapter, the authors point out that productivity rose due to an increase in specialization and the introduction of special technologies and farm machinery, when farmers turned to cultivation of monocultures, which is hard on land resources compared to crop rotation. This led to environmental deterioration and the displacement of people from the land, leading to the development of slums in the Third World. Contemporary farming has transformed the world's landscape and endangered the existence of life on this planet, as related in the last chapter. The authors predict that farmers, especially those in developing countries, need to be more educated to increase worldwide food production by 2% to 3% per annum, as well as to stabilize the world's population. The information cited is well documented, and a bibliography is included. The authors have produced a useful, engaging, and informative overview of environmental concerns. This book will be useful for an undergraduate interdisciplinary course in biology, agriculture, and ecology."
—Reviewed by N. N. Raghuvir in *Science Books & Films*, 30/7 (August/September 1994), p. 165.

Contents

Early Food Acquisition

From Hunter-Gatherers to Farmers

Early Agriculture

Domesticating Plants

The Rise of Civilization

Agriculture Spreads to Europe

The Medieval Farm

Sugarcane and Industrial Agriculture

A New Kind of Farm

Contemporary Farming

The Future of Food

Science for All Americans / CHAPTER 9: THE MATHEMATICAL WORLD

Archimedes' Revenge: The Joys and Perils of Mathematics	Paul Hoffman	Fawcett Crest	1988
Discovering Mathematics: The Art of Investigation	A. Gardiner	Oxford University Press	1987
An Equation That Changed the World: Newton, Einstein, and the Theory of Relativity	Harald Fritzsch and translated by Karin Heusch	University of Chicago Press	1994
From Zero to Infinity, 4th ed.	Constance Reid	Mathematical Association of America	1992
How to Lie with Statistics	Darrell Huff	W. W. Norton	1954
How to Solve It	George Polya	Princeton University Press	1945
How We Know: An Exploration of the Scientific Process	Martin Goldstein and Inge F. Goldstein	Plenum	1978
Innumeracy: Mathematical Illiteracy and Its Consequences	John Allen Paulos	Hill & Wang	1988
Lady Luck	Warren Weaver	Dover	1963
The Mathematical Universe	William Dunham	John Wiley & Sons	1994
On the Shoulders of Giants: New Approaches to Numeracy	National Research Council	National Academy Press	1990
One Two Three...Infinity	George Gamow	Dover	1947
Recent Revolutions in Mathematics	Albert Stwertka	Franklin Watts Inc.	1987
The Refrigerator and the Universe: Understanding the Laws of Energy	Martin Goldstein and Inge F. Goldstein	Harvard University Press	1993
Statistics Concepts and Controversies	David S. Moore	W. H. Freeman and Co.	1985
Time Travel and Other Mathematical Bewilderments	Martin Gardner	W. H. Freeman and Co.	1988
The World of Mathematics	James R. Newman	Simon & Schuster	1956

SFAA link: 9D

On the Shoulders of Giants:
New Approaches
to Numeracy

National Research Council
(Illus.)

National Academy Press

1990

232 pp.

$17.95

0-309-04234-8

Index

C, T, GA **

"This report, commissioned by the National Research Council, is intended to stimulate creative approaches to mathematics curricula in the next century. Unlike the David Report (National Research Council, 1984), which examined the state of mathematics research, support for research, and the rapport between mathematics and various areas of application, this report is concerned primarily with the instruction of mathematics in the elementary and secondary grades. It is addressed to teachers, university faculty, the public, and those in positions of public trust. The volume is organized into six essays written by separate authors. These are "Pattern" by Lynn Arthur Steen, "Dimension" by Thomas F. Banchoff, "Quantity" by James T. Fey, "Uncertainty" by David S. Moore, "Shape" by Marjorie Senechal, and "Change" by Ian Stewart. The volume addresses a number of aspects on the problem of innumeracy—the public's general lack of understanding that mathematics is a dynamic discipline and the importance of earlier learning as a prelude to subsequent knowledge both in a historical sense (hence the allusion in the title) and as a mode of instruction so important for mathematics. Most of the examples and illustrations can be understood with a minimum of mathematics, but they also provide a powerful motivation for overhauling and upgrading mathematics instruction. There is a differentiation made between understanding and doing mathematics, and between mathematics as a language and as an exploratory science without deprecating any of these aspects of mathematics. A teacher at almost any level will find some useful ideas and projects as well as a broader grasp of what mathematics is about. The problem that is left unresolved (perhaps because it does not have a simple solution) is not whether to teach fundamentals, but rather which fundamentals to teach. Separately and collectively these presentations show that a massive reorganization of mathematics instruction is required in order for students in U.S. schools to reach the standards we all claim we want them to meet. The traditional separation of arithmetic, algebra, geometry, and calculus is not adequate for the current state of mathematics."
—Reviewed by Donald E. Myers in *Science Books & Films*, 27/1 (January/February 1991), p. 5-6.

Contents

Pattern / Lynn Arthur Steen
Dimension / Thomas F. Banchoff
Quantity / James T. Fey
Uncertainty / David S. Moore
Shape / Marjorie Senechal
Change / Ian Stewart

Science for All Americans / Chapter 10: Historical Perspectives

The Beak of the Finch: A Story of Evolution in Our Time	Jonathan Weiner	Knopf	1994
Blueprints: Solving the Mystery of Evolution	Maitland A. Edey and Donald C. Johanson	Little, Brown	1989
Charles Darwin: Evolution of a Naturalist	Richard Milner	Facts on File	1994
The Control Revolution: Technical and Economic Origins of the Information Society	James R. Beniger	Harvard University Press	1986
The Day the Sun Rose Twice	Ferenc Morton Szasz	University of New Mexico Press	1984
Discovery, Innovation and Risk: Case Studies in Science and Technology	Newton H. Copp and Andrew W. Zanella	MIT Press	1993
Discovery of Time	Stephen Toulmin and June Goodfield	Harper & Row	1965
Engines of Change: The American Industrial Revolution, 1790-1860	Brooke Hindle and Steven Lubar	Smithsonian Institution Press	1986
Ever Since Darwin: Reflections in Natural History	Stephen Jay Gould	W. W. Norton	1977
The Fabric of the Heavens	Stephen Toulmin and June Goodfield	Harper & Row	1961
The History of Modern Science: A Guide to the Second Scientific Revolution, 1800-1950	Stephen G. Brush	Iowa State University Press	1988
The History of Science from 1895 to 1945	Ray Spangenburg and Diane K. Moser	Facts on File	1994
The History of Science from the Ancient Greeks to the Scientific Revolution	Ray Spangenburg and Diane K. Moser	Facts on File	1993
The History of Science in the Eighteenth Century	Ray Spangenburg and Diane K. Moser	Facts on File	1993
The History of Science in the Nineteenth Century	Ray Spangenburg and Diane K. Moser	Facts on File	1994
A History of the Sciences	Stephen F. Mason	Collier	1962
The Major Achievements of Science: The Development of Science from Ancient Times to the Present	A. E. E. McKenzie	Iowa State University Press	1988
Mammal Evolution: An Illustrated Guide	R. J. G. Savage	Facts on File	1986
Microbe Hunters	Paul de Kruif	Harcourt Brace Jovanovich	1926
New Theories on the Origins of the Human Race	Christopher Lampton	Franklin Watts Inc.	1989
Nuclear Choices: A Citizen's Guide to Nuclear Technology	Richard Wolfson	MIT Press	1991

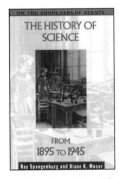

"The first part of this wonderfully readable science survey covers developments in the physical sciences in the first half of the twentieth century, a period that saw changes in ideas about the structure of the atom, the nature of light waves, outer space, and relativity. The second part of the book deals with the life sciences: microbiology, biochemistry, genetics, and archaeology, with each science linked with the lives of scientists studying it. Black-and-white photos are numerous, and anecdotes about the personalities of the scholars add interest. The authors present the scientists and their accomplishments against a backdrop of two world wars and include an epilogue that looks into the second half of this century. An appendix on the scientific method, a chronology, a brief glossary, and a bibliography are appended."
—Reviewed by Susan DeRonne in *Booklist*, (September 1994).

SFAA link: 10C 10G 10H 10I

Spangenburg, Ray, and Diane K. Moser
The History of Science from 1895 to 1945
Facts on File
1994
192 pp.
$18.95
0-8160-2742-0
Bibliography; Glossary
Index

Contents

An Overview: 1895-1945

Part One: The Physical Sciences

The New Atom: From X Rays to the Nucleus

The New Universe, Part One: Einstein and Relativity

The New Universe, Part Two: The Quantum Surprise

New Observations of the Universe

The Atom Split Asunder: Science and the Bomb

Part Two: The Life Sciences

The Growth of Microbiology and Biochemistry

Pursuing the Trails of Genetics and Heredity

In Search of Ancient Humans

Science for All Americans / Chapter ii: Common Themes

Charles Darwin: Evolution of a Naturalist	Richard Milner	Facts on File	1994
The Common Sense of Science	Jacob Bronowski	Harvard University Press	1978
The Control Revolution: Technical and Economic Origins of the Information Society	James R. Beniger	Harvard University Press	1986
Cycles of Nature: An Introduction to Biological Rhythms	Andrew Ahlgren and Franz Halberg	National Science Teachers Association	1990
Diatoms to Dinosaurs: The Size and Scale of Living Things	Chris McGowan	Island Press	1994
Engineering and the Mind's Eye	Eugene S. Ferguson	MIT Press	1992
Engines of Change: The American Industrial Revolution, 1790–1860	Brooke Hindle and Steven Lubar	Smithsonian Institution Press	1986
Knowledge and Wonder—The Natural World as Man Knows It	Victor F. Weisskopf	Doubleday	1979
Man on Earth	John Reader	University of Texas Press	1988
Metaman: The Merging of Humans and Machines into a Global Superorganism	Gregory Stock	Simon & Schuster	1993
Our Natural Resources and Their Conservation, 7th ed.	Harry Kircher, Donald Wallace, and Dorothy Gore	Interstate Publishers	1991
Powers of Ten: A Book about the Relative Size of Things in the Universe and the Effect of Adding Another Zero	Philip and Phylis Morrison and the office of Charles and Ray Eames	*Scientific American* Library	1994
The Scientific Attitude, 2nd ed.	Frederick Grinnell	The Guilford Press	1992
The Search for Solutions	Horace Freeland Judson	Holt, Rinehart and Winston	1980
On the Shoulders of Giants: New Approaches to Numeracy	National Research Council	National Academy Press	1990
Supercomputing and the Transformation of Science	William J. Kaufmann III and Larry L. Smarr	*Scientific American* Library	1993
The World of Mathematics	James R. Newman	Simon & Schuster	1956

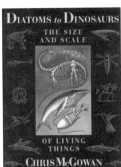

SFAA link: 5A 5F 11D

McGowan, Chris

Diatoms to Dinosaurs:
The Size and Scale
of Living Things
(Illus. by Julian Mulock)
Island Press
1994
272 pp.
$24.95
1-55963-304-2
Glossary; Index
C, T, GA **

"Chris McGowan's fascinating text, *Diatoms to Dinosaurs*, melds together many facets of evolutionary biology to explain the complex interrelationship between animal body size and survival. In an extremely accessible and anecdotal style reminiscent of that employed by Stephen Jay Gould, McGowan explores the effective relationship between, on the one hand, animal body size and, on the other, metabolism, strength, and life expectancy throughout evolutionary history. In so doing, he utilizes numerous animal ecology models, literally from diatoms to dinosaurs and on to vertebrates. His treatment of the relation of animal brain size to intellectual ability, as well as the successful application of physical laws and evolved structural components such as feathers and fins to an animal's flight dynamics and swimming abilities, is particularly mesmerizing. The numerous illustrations and drawings superbly complement the riveting text. I highly recommend this excellent book equally to the armchair popular science aficionado and the undergraduate or graduate-level evolutionary biologist or ecologist as a satisfying and enjoyable read."
—Reviewed by Robert R. J. Grispino in *Science Books & Films*, 31/2 (March 1995), p. 39.

Contents

The Scale of Life

Keep the Home Fires Burning

Pumping Iron

From Cradle to Grave

Giants—Modern and Ancient

Brains: From the Massive to the Minute

Drag in the Material World

High Fliers

Tiffany Wings and Kite Strings

Drifting with the Tide: Life in the Plankton

Life in the Fast Lane

Science for All Americans / CHAPTER 12: HABITS OF MIND

How to Lie with Statistics	Darrell Huff	W. W. Norton	1954
Innumeracy: Mathematical Illiteracy and Its Consequences	John Allen Paulos	Hill & Wang	1988
Inventors at Work: Interviews with 16 Notable American Inventors	Kenneth A. Brown	Tempus	1988
A Mathematician Reads the Newspaper	John Allen Paulos	Basic Books	1995
Pasteur and Modern Science	Rene Dubos	Science Tech	1988
A Physicist on Madison Avenue	Tony Rothman	Princeton University Press	1991
The Refrigerator and the Universe: Understanding the Laws of Energy	Martin Goldstein and Inge F. Goldstein	Harvard University Press	1993
The Science Gap: Dispelling the Myths and Understanding the Reality of Science	Milton A. Rothman	Prometheus Books	1992
Statistics Concepts, and Controversies	David S. Moore	W. H. Freeman	1979
You Know What They Say... The Truth about Popular Beliefs	Alfie Kohn	HarperCollins	1990

SFAA link: 1A 1B 12E

Rothman, Milton A.

*The Science Gap:
Dispelling the Myths and
Understanding the
Reality of Science*

Prometheus Books

1992

254 pp.

$24.95

0-87975-710-8

Index

YA-T, GA **

"The subtitle of Rothman's book describes the content of this extraordinary volume. The author, a professor of physics and a research physicist at the Princeton Plasma Physics Laboratory, is in an excellent position, at the frontier of science and engineering, to speak with great authority. This work is directed to the educated lay reader. It has no equations and no technical language; yet so clear is the exposition, that both the novice and the self-styled expert can benefit from its demolition of the widespread fallacies concerning both the physical universe and the community of scientists who observe, explore, and explain it. In the terminology of the trade, Rothman is a "critical realist." He defends this position with relentless logic. For a generation that has grown up with *Star Trek*, an immense library of science fiction, and media that entertain rather than educate, this book will be harsh but desperately needed therapy. The myths demolished range from such old chestnuts about perpetual motion machines to more contemporary mythologies concerning the way scientists reach communal certainty. Those who believe in UFOs, ghosts, telepathy, and other arcana of *The Twilight Zone* will find no comfort here. Of immediate contemporary significance is the author's careful analysis of the cold fusion hoopla. The National Science Foundation would be well advised to put a copy of this book in every high school and college."

—Reviewed by Robert G. Colodny in *Science Books & Films*, 28/5 (June/July 1992), p. 140.

Contents

MYTH 1: "Nothing exists until it is observed."

MYTH 2: "Nothing is known for sure."

MYTH 3: "Nothing is impossible."

MYTH 4: "Whatever we think we know now is likely to be overturned in the future."

MYTH 5: "Advanced civilizations of the future will have the use of forces unknown to us at present."

MYTH 6: "Advanced civilizations on other planets possess great forces unavailable to us on earth."

MYTH 7: "There are more things in heaven and earth, Horatio, than are dreamt of in your philosophy."

MYTH 8: "Scientists don't have any imagination."

MYTH 9: "Scientists create theories by intuition."

MYTH 10: "All theories are equal."

MYTH 11: "All scientists are objective."

MYTH 12: "Scientists are always making false predictions."

MYTH 13: "All problems can be solved by computer modeling."

MYTH 14: "More technology will solve all problems."

MYTH 15: Myths about Reductionism

MYTH 16: "Myths are just harmless fun and good for the soul."

FUTURE VERSIONS

Many excellent books related to *Science for All Americans* topics have no doubt evaded our initial search process. In preparation for the next version of *Resources for Science Literacy*, Project 2061 will extend its search for trade books and will also seek out journal articles and other similar materials that meet the selection criteria. In the process, it will identify persistent gaps in the literature and perhaps encourage publishers and authors to fill them. Users are encouraged to contribute their own lists of books and articles and to suggest other ways to improve the database.

Send suggestions to:
Project 2061
American Association for the Advancement of Science
1333 H Street, NW
P.O. Box 34446
Washington, D.C. 20005
FAX: 202/842-5196
Electronic Mail: project2061@aaas.org (please identify subject as "books")

ALSO SEE

For more detailed instructions on using the trade-books database, please refer to Chapter 7: Using the *Resources for Science Literacy* CD-ROM on page 107 of this volume.

ALBERT PALEY, *Mystery Table, 1993*.

Chapter *3* COGNITIVE RESEARCH

Many teachers have little access to the latest education research or experience applying it to instruction. Most would agree, however, with the following assertion in *Benchmarks for Science Literacy*:

> Overestimation of what students can learn at a given age results in student frustration, lack of confidence, and unproductive learning strategies, such as memorization without understanding. Underestimating what students can learn results in boredom, overconfidence, poor study habits, and a needlessly diluted education. So it is important to make decisions about what to expect of students and when on the basis of as much good information as possible.

Because understanding the research on how students learn science concepts is so important to promoting science literacy goals, *Benchmarks for Science Literacy* included in its Chapter 15: The Research Base a survey of the research that influenced the content and grade-level placement of benchmark ideas. However, it did not include any articles, reports, or books specifically written for teachers.

The cognitive research database of *Resources for Science Literacy: Professional Development* has been created to supplement *Benchmarks* Chapter 15 with references to additional reading that teachers can pursue out of their own personal interest or as part of their in-service/pre-service education. The database includes descriptions of over 80 articles and reports, 17 books, and four videos related to student learning of concepts from *Science for All Americans (SFAA)* and *Benchmarks for Science Literacy*. Educators who want to take into account appropriate research findings as they make decisions about curriculum, instruction, and assessment will be able to do so through a variety of search and sorting functions.

NOTES ON THE COGNITIVE RESEARCH DATABASE ON CD-ROM

The cognitive research database includes four sections:

- Descriptions of articles and published reports, appropriate for teacher audiences, that describe and discuss student ideas in science.
- Descriptions of books and collections of articles, appropriate for use in pre-service or in-service teacher education, that describe and discuss student ideas in science.

- Descriptions of videos that illustrate difficulties students have in learning science.
- *Benchmarks* Chapter 15: The Research Base, and its accompanying bibliography of more than 300 references to cognitive research literature.

Citations for each article, book, and report in the database provide full bibliographic information, a short description or abstract of the work's content and its targeted audience, links to *SFAA* chapters and sections, and the grade level focus of the work. Published reviews of books are also included. For each video, the database provides a brief description of the video's content along with its running time, links to *SFAA,* and availability.

All of the articles, books, reports, and videos included in the cognitive research database meet two criteria:

- They are directly relevant to concepts from *SFAA* and *Benchmarks for Science Literacy.*
- They are of high quality and accurately represent the research.

In the case of articles, the emphasis is on identifying research published in journals that are readily available to teachers, such as *The Physics Teacher, The Science Teacher, Science and Children, The American Biology Teacher, Mathematics Teacher,* and *Arithmetic Teacher.*

Examples from the CD-ROM

All of the research materials—articles, books, reports, and videos—cited on the *Professional Development* CD-ROM are also listed in the following pages, with examples of each type of research citation available. In addition to providing full bibliographic information and an abstract of the research, each citation includes the grade range to which the research applies and the *SFAA* chapters and sections to which the research is linked.

As an example of the kinds of research referred to in Chapter 15 of *Benchmarks for Science Literacy,* all of the citations relevant to the Scientific World View section (1A) of Chapter 1: The Nature of Science, in *SFAA* are presented in the example taken from the CD-ROM.

ARTICLES

Action Research: Strategies for Learning Subtraction Facts	Carol A. Thornton and Paula J. Smith	*Arithmetic Teacher*	April 1988
Children's Conceptual Understanding of Forces and Equilibrium	C. Terry and others	*Physics Education*	July 1985
Children's Difficulties in Subtraction: Some Causes and Cures	Arthur J. Baroody	*Arithmetic Teacher*	November 1984
Children's Dynamics	Roger Osborne	*Physics Teacher*	November 1984
Concepts in Force and Motion	Nanjundiah Sadanand and Joseph Kess	*Physics Teacher*	November 1990
Connecting Research to Teaching	Kathleen Cramer and Thomas Post	*Mathematics Teacher*	May 1993
Connecting Research to Teaching: Probability and Statistics	J. Michael Shaughnessy	*Mathematics Teacher*	March 1993
Constructing Meaning for the Concept of Equation	Nicolas Herscovics and Carolyn Kieran	*Mathematics Teacher*	November 1980
A Constructivist Approach to Astronomy in the National Curriculum	John Baxter	*Physics Education*	January 1991
The Continuing Calculator Controversy	Thomas Dick	*Arithmetic Teacher*	April 1988
Decimal Notation: An Important Research Finding	Anne S. Grossman	*Arithmetic Teacher*	May 1983
Decimals: Results and Implications from National Assessment	Thomas P. Carpenter and others	*Arithmetic Teacher*	April 1981
The Everyday Perspective and Exceedingly Unobvious Meaning	Charles R. Ault, Jr.	*Journal of Geological Education*	March 1984
Force Concept Inventory	David Hestenes and others	*Physics Teacher*	March 1992
Fractions: In Search of Meaning	Mary Lou Witherspoon	*Arithmetic Teacher*	April 1993
Fractions: Results and Implications from National Assessment	Thomas R. Post	*Arithmetic Teacher*	May 1981
Getting the Facts Straight	Jim Minstrell	*Science Teacher*	January 1983
Gravity—Don't Take It for Granted	Michael D. Watts	*Physics Education*	May 1982
How Well Do Students Write Geometry Proofs?	Sharon L. Senk	*Mathematics Teacher*	September 1985
Let Us Go Back to Nature Study	Dorothy L. Gabel	*Journal of Chemical Education*	September 1989
Little into Big Is the Way It Always Is	Anna O. Graeber and Kay M. Baker	*Arithmetic Teacher*	April 1992

ARTICLES

Material Cycles in Nature: A New Approach to Teaching Photosynthesis in Junior High School	Yehudit Eisen and Ruth Stavy	*American Biology Teacher*	September 1992
Misconceptions and the Qualitative Method	Dori Ridgeway	*Science Teacher*	September 1988
The Museum as Science Teacher	Charles R. Ault, Jr.	*Science and Children*	November–December 1987
Research Gives Calculators a Green Light	Ray Hembree	*Arithmetic Teacher*	September 1986
Research into Practice: Adding by Counting On with One-Handed Finger Patterns	Karen C. Fuson	*Arithmetic Teacher*	September 1987
Research into Practice: Children's Multiplication	Kurt Kilion and Leslie Steffe	*Arithmetic Teacher*	September 1989
Research into Practice: Children's Strategies in Ordering Rational Numbers	Thomas Post and Kathleen Cramer	*Arithmetic Teacher*	October 1987
Research into Practice: Children's Understanding of Zero and Infinity	Marilyn Montague Wheeler	*Arithmetic Teacher*	November 1987
Research into Practice: Developing Understanding of Computational Estimation	Judith Sowder	*Arithmetic Teacher*	January 1989
Research into Practice: Helping to Make the Transition to Algebra	Carolyn Kieran	*Arithmetic Teacher*	March 1991
Research into Practice: Integrating Assessment and Instruction	Megan M. Loef, Deborah A. Carey, Thomas P. Carpenter, and Elizabeth Fennema	*Arithmetic Teacher*	November 1988
Research into Practice: Misconceptions about Multiplication and Division	Anna O. Graeber	*Arithmetic Teacher*	March 1993
Research into Practice: Number Sense and Nonsense	Zvia Markovits and others	*Arithmetic Teacher*	February 1989
Research into Practice: Place Value and Addition and Subtraction	Diane Wearne and James Hiebert	*Arithmetic Teacher*	January 1994
Research into Practice: Place Value for Tens and Ones	Joseph N. Payne	*Arithmetic Teacher*	February 1983
Research into Practice: The Problem of Fractions in the Elementary School	Leslie P. Steffe and John Olive	*Arithmetic Teacher*	May 1991
Research into Practice: Similarity in the Middle Grades	Glenda Lappan and Ruhama Even	*Arithmetic Teacher*	May 1988
Research into Practice: Spatial Abilities	Douglas T. Owens	*Arithmetic Teacher*	February 1990
Research into Practice: Story Problems and Students' Strategies	Larry Sowder	*Arithmetic Teacher*	May 1989
Research into Practice: Students' Use of Symbols	Deborah A. Carey	*Arithmetic Teacher*	November 1992

Articles

Research into Practice: Using Spatial Imagery in Geometric Reasoning	Michael T. Battista and Douglas H. Clements	*Arithmetic Teacher*	November 1991
Research Report: Addition and Subtraction—Processes and Problems	Marilyn N. Suydam	*Arithmetic Teacher*	December 1985
Research Report: Decimal Fractions	James Hiebert	*Arithmetic Teacher*	March 1987
Research Report: Estimation and Computation	Marilyn N. Suydam	*Arithmetic Teacher*	December 1984
Research Report: Learning Probability Concepts in Elementary School Mathematics	Albert P. Shulte	*Arithmetic Teacher*	January 1987
Research Report: What Are Calculators Good For?	Marilyn N. Sudam	*Arithmetic Teacher*	February 1987
Results of the Fourth NAEP Assessment of Mathematics: Measurement, Geometry, Data Interpretation, Attitudes, and Other Topics	Vicky L. Kouba and others	*Arithmetic Teacher*	May 1988
Secondary School Results for the Fourth NAEP Mathematics Assessment: Algebra, Geometry, Mathematical Methods, and Attitudes	Catherine A. Brown and others	*Mathematics Teacher*	May 1988
Secondary School Results for the Fourth NAEP Mathematics Assessment: Discrete Mathematics, Data Organization and Interpretation, Measurement, Number and Operations	Catherine A. Brown and others	*Mathematics Teacher*	April 1988
Secondary Students' Conceptions of the Conduction of Heat: Bringing Together Scientific and Personal Views	Elizabeth Engel Clough and Rosalind Driver	*Physics Education*	July 1985
The Shape of Instruction in Geometry: Some Highlights from Research	Marilyn N. Suydam	*Mathematics Teacher*	September 1985
Some Misconceptions Concerning the Concept of Variables	Peter Rosnick	*Mathematics Teacher*	September 1981
Spadework Prior to Deduction in Geometry	J. Michael Shaughnessy and William F. Burger	*Mathematics Teacher*	September 1985
Students' Conceptions of Ideas in Mechanics	John K. Gilbert and others	*Physics Education*	March 1982
Students' Concepts of Force: The Importance of Understanding Newton's Third Law	David E. Brown	*Physics Education*	November 1989
Students' Understanding of Photosynthesis	Yehudit Eisen and Ruth Stavy	*American Biology Teacher*	April 1988
Students' Use of the Principle of Energy Conservation in Problem Situations	Rosalind Driver and Lynda Warrington	*Physics Education*	July 1985
A Survey of Some Children's Ideas about Force	D. M. Watts and A. Zylbersztajn	*Physics Education*	November 1981

ARTICLES

Teacher Predictions versus Actual Student Gains	Alan Lightman and Philip Sadler	*Physics Teacher*	March 1993
Teaching for Conceptual Change: Confronting Children's Experience	Bruce Watson and Richard Konicek	*Phi Delta Kappan*	May 1990
Understanding the Particulate Nature of Matter	Dorothy L. Gabel and others	*Journal of Chemical Education*	August 1987
Units of Measure: Results and Implications from National Assessment	James Hiebert	*Arithmetic Teacher*	February 1981
What Are the Chances of Your Students Knowing Probability?	Thomas P. Carpenter and others	*Mathematics Teacher*	May 1981
What Are These Things Called Variables?	Sigrid Wagner	*Mathematics Teacher*	October 1983
What Research Says	Mark Driscoll	*Arithmetic Teacher*	February 1984
What Research Says: The Cardiovascular System: Children's Conceptions and Misconceptions	Mary W. Arnaudin and Joel J. Mintzes	*Science and Children*	February 1986
When Students Don't Know They Don't Know	Janet F. Eaton and others	*Science and Children*	April 1983
Which Comes First—Length, Area, or Volume?	Kathleen Hart	*Arithmetic Teacher*	February 1981
Why Are There No Dinosaurs in Oklahoma?	John W. Renner and others	*Science Teacher*	December 1981
Why Do Some Children Have Trouble Learning Measurement Concepts?	James Hiebert	*Arithmetic Teacher*	March 1984

EXAMPLES OF ARTICLE CITATIONS

Connecting Research to Teaching: Probability and Statistics
J. Michael Shaughnessy
Mathematics Teacher
Volume 86,
Number 3
March 1993
pp. 244-48
Grade focus: 6-12
SFAA link: 9D, 12B

Presents research findings related to students' intuitive ideas about the concepts of chance to inform teachers how students form their concepts of probability and statistics. Discusses adolescents' conceptions of uncertainty, judgmental heuristics in making estimates of event likelihood, the conjunction fallacy, the outcome approach, attempts to change probabilistic beliefs, and teaching methods. (22 references)

Available from University Microfilms, International

Force Concept Inventory
David Hestenes
and others
Physics Teacher
Volume 30,
Number 3
March 1992
pp. 141-58
Grade focus: 9-12
SFAA link: 4F, 4G

Reports the rationale, design, validation, and uses of the "Force Concept Inventory," an instrument to assess the students' beliefs on force. Includes results and implications of two studies that compared the inventory with the "Mechanics Baseline." Includes a copy of the instrument.

Available from University Microfilms, International

Material Cycles in Nature:
A New Approach to
Teaching Photosynthesis
in Junior High School

Yehudit Eisen
and Ruth Stavy
American Biology Teacher
Volume 54,
Number 6
September 1992
pp. 339-42
Grade focus: 6-8
SFAA link: 5E

This paper describes the current approach used in Israel for teaching photosynthesis to junior high school students. Authors identify two types of problems, psychological and instructional, that children experience in understanding photosynthesis. Presents a foundation and sequence for a new approach to teaching photosynthesis. (18 references)

Available from University Microfilms, International

Books

Analysis of Arithmetic for Mathematics Teaching	Gaea Leinhardt, Ralph Putnam, and Rosemary A. Hattrup	Lawrence Erlbaum	1992
The Child's Construction of Economics	Anna Emilia Berti and Anna Silvia Bombi	Cambridge University Press	1981
Elementary School Social Studies: Research as a Guide to Practice	Virginia A. Atwood	National Council for the Social Studies	1991
Handbook of Research on Mathematics Teaching and Learning	Doug Grouws	Macmillan	1992
Handbook of Research on Science Teaching and Learning	Dorothy L. Gabel	Macmillan	1994
Handbook of Research on Social Studies Teaching and Learning	James P. Shaver	Macmillan	1991
"The Having of Wonderful Ideas" & Other Essays on Teaching & Learning	Eleanor Duckworth	Teachers College Press	1987
Learning in Science: The Implications of Children's Science	Roger Osborne and Peter Freyberg	Heinemann	1985
Making Sense of Secondary Science: Research into Children's Ideas	Rosalind Driver, Ann Squires, Peter Rushworth, and Valerie Wood-Robinson	Routledge	1994
The Psychology of Learning Science	Shawn M. Glynn, Russell H. Yeany, and Bruce K. Britton	Lawrence Erlbaum	1991
The Pupil as Scientist?	Rosalind Driver	Open University Press	1983
Research Ideas for the Classroom: Early Childhood Mathematics	Robert J. Jensen	Macmillan	1993
Research Ideas for the Classroom: High School Mathematics	Patricia S. Wilson	Macmillan	1993
Research Ideas for the Classroom: Middle Grades Mathematics	Douglas T. Owens	Macmillan	1993
What Research Says to the Science Teacher: The Process of Knowing	Mary Budd Rowe	National Science Teachers Association	1990

Making Sense of
Secondary Science:
Research into
Children's Ideas
Rosalind Driver,
Ann Squires,
Peter Rushworth, and
Valerie Wood Robinson
Routledge
1994
0-415-09767-3;
0-415-09765-7 (paper)
Grade focus: 3-12
SFAA link: 4, 5, 6B,
6C, 6E

Making Sense of Secondary Science provides a comprehensive and concise review of student ideas in science in most topics that are included in *Science for All Americans* Chapter 4: The Physical Setting, Chapter 5: The Living Environment, and Chapter 6: The Human Organism.

The research findings are arranged in three sections: Life and Living Processes, Materials and Their Properties, and Physical Processes. Each section includes student ideas on topics such as living things, nutrition, growth, and ecosystems. For each topic, there is an extensive list of references so readers can refer in more detail to the studies themselves.

Although the title suggests that the primary audience is secondary school teachers, the book will be of interest and value to science teachers at all grade levels who want to deepen their understanding of how their students think.

What Research Says
to the Science Teacher:
The Process of Knowing
Mary Budd Rowe
National Science
Teachers Association
1990
0-87355-093-5
Grade focus: 3-12
SFAA link: 4D, 4F

What Research Says to the Science Teacher: The Process of Knowing examines the implications of cognitive science research for improvement of science education. It is part of an ongoing series published by the National Science Teachers Association.

Two chapters include summaries of research on students' ideas about topics addressed in *Science for All Americans*. Chapter 5: Graphs, Graphing and Graphers identifies difficulties students have in constructing and interpreting graphs. Chapter 6: The Uncommon Common Sense of Science includes brief summaries of research findings on students' ideas about topics such as forces and motion, weight and density, the particulate nature of matter, and conservation of matter.

The book's Chapter 3: Implications of Teachers' Conceptions of Science Teaching and Learning examines teachers' ideas about science teaching and learning and discusses their consequences for teachers' professional development. The volume should be of interest and value to teachers and to those who are involved in the professional development of teachers.

REPORTS

Aspects of Secondary Students' Understanding of Energy: Summary Report	Angela Brook and Rosalind Driver	University of Leeds	1984
Aspects of Secondary Students' Understanding of Heat: Summary Report	Angela Brook, Hazel Briggs, Beverly Bell, and Rosalind Driver	University of Leeds	1984
Aspects of Secondary Students' Understanding of Particles	Angela Brook, Hazel Briggs, and Beverly Bell	University of Leeds	1983
Aspects of Secondary Students' Understanding of Plant Nutrition: Summary Report	Beverly Bell	University of Leeds	1984
The Earth in Space	Jonathan Osborne, Pam Wadsworth, Paul Black, and John Meadows	Liverpool University Press	1994
Growth	Terry Russell and Dorothy Watt	Liverpool University Press	1990
Light	Jonathan Osborne, Paul Black, Maureen Smith, and John Meadows	Liverpool University Press	1990
Materials	Terry Russell, Ken Longden, and Linda McGuigan	Liverpool University Press	1991
Processes of Life	Jonathan Osborne, Pam Wadsworth, and Paul Black	Liverpool University Press	1992
Progression in Science: The Development of Pupils' Understanding of Physical Characteristics of Air, Across the Age Range 5-16 Years	Angela Brook and Rosalind Driver in collaboration with Dudley Hind	University of Leeds	1984
Rocks, Soil, and Weather	Terry Russell, Derek Bell, Ken Longden, and Linda McGuigan	Liverpool University Press	1993
Sound	Dorothy Watt and Terry Russell	Liverpool University Press	1990

Rocks, Soil, and Weather

Terry Russell, Derek Bell, Ken Longden, and Linda McGuigan

Liverpool University Press[1]

1993

0-85323-497-3

Price: £7.00

Grade focus: K-2, 3-5

SFAA link: 4

This report is one of a series developed by the Primary SPACE Project (Science Processes and Concept Exploration), based jointly at the Department of Education, University of Liverpool, and the Center for Educational Studies, King's College London, UK. The series describes research conducted by researchers and teachers to understand the ideas which elementary school children have in particular science concept areas and how students modify their ideas as the result of brief teaching interventions.[2]

This report contains: (1) a summary of previous research into children's ideas related to soil, rocks, the earth's structure, and weather; (2) a description of student ideas about rocks, soil, and the weather, before they received any formal experiences relating to these concepts; (3) a description of the teaching intervention that was intended to encourage students to develop their ideas; (4) a description of student ideas after the teaching intervention (note: student's ideas of weather were elicited only before the intervention); and (5) a summary of the research findings and implications for curriculum and instruction. Appendices include brief descriptions of the interview tasks that were used to explore student ideas, and of the activities within the teaching interventions. Excerpts from student interviews and samples of student drawings and written work appear throughout the report.

Some findings include: After the intervention, increased proportions of students thought of soil as a complex mixture, rather than as a homogenous material; recognized that soil contains different sized particles of the same inorganic constituent and that it contains living organisms; recognized that soil can be transformed and mentioned the sea and rain as agents of transformation; and drew layers of rock when asked about what may lie beneath the surface of the earth.

Other titles in the series include: *The Earth in Space, Evaporation and Condensation, Growth, Light, Materials, Processes of Life,* and *Sound.*

---

[1] Available through: Burston Distribution Services, Unit 2A, Newbridge Trading Estate, Newbridge Close, off Whitby Road, Bristol BS4 4AX, UK; call: 011 441179 724248, fax: 011 441179 711056.

[2] The report includes references to stages and attainment targets of the British national curriculum that may be confusing to U.S. readers. The terminology used to characterize different age spans may also be confusing: The term *infants* refers to children from five to seven, *lower juniors* to children from seven to nine, and *upper juniors* to children from nine to eleven.

VIDEOS

Programs from The Private Universe Workshop Series	The following videos have been selected from a nine-program series produced by the Harvard-Smithsonian Center for Astrophysics. Expanding on the concept presented in the original *Private Universe* film (see below), each video focuses on a single theme and content area and uses specific examples to show how students' preconceived ideas can create barriers to learning. All of the videos are available from Annenberg/CPB Math and Science Collection, P.O. Box 2345, S. Burlington, VT 05407-2345 (1-800-965-7373).
A Private Universe	This 18-minute video demonstrates that even after many years of "the best education that money can buy," many students—including Harvard graduates—had inaccurate and inconsistent ideas about basic science.
Program 1. Astronomy: Eliciting Student Ideas	This 90-minute video includes student interviews about astronomy, discussion of issues the students raise, and strategies for effective teaching. Interviews with Harvard graduates and faculty show that they are confused about what causes the seasons. We also see Heather, an articulate, intelligent high-school student, who has a great many ideas about astronomy. Interviews with Heather both before and after her classroom lessons about astronomy reveal that she has learned much but is still confused about some key aspects of the subject. Although some of Heather's ideas after instruction are solid, others are inconsistent with accepted scientific ideas. Some of her ideas stubbornly resist change, both in the classroom and during on-camera challenges.

Program 2. Biology: Lessons Pulled from Thin Air

This 90-minute video includes student interviews about the source of a log's mass, discussion of issues students raise, and strategies for effective teaching. One interview, with Harvard and MIT graduates, demonstrates that even college graduates hold misconceptions about basic concepts in science. When 21 graduates from Harvard and MIT were shown a log and were asked to explain where its mass and that of the rest of the tree came from, none mentioned the air as a source of the log's mass. Some graduates were surprised, and some even disagreed, when the interviewer suggested that the mass of the tree came mostly from the carbon dioxide in the air. Extensive interviews with a middle-school student show similar misconceptions and illustrate how some student ideas are quite resistant to change.

Program 4. Chemistry: A Schoolhouse with No Foundation

This 90-minute video includes student interviews about the particulate nature of matter, discussion of issues the students raise, and strategies for effective teaching. In one interview Jamie, a gifted 8th-grader, thinks about some phenomena that relate to the particulate model of matter. Jamie is presented with two tasks. In the first, she is presented with a syringe filled with air. The air in the syringe is compressed ("squashed"), and Jamie is asked to draw a diagram of the air inside the syringe before and after it is squashed. In the second task, Jamie is presented with a closed flask containing air. She is asked to describe what the air would look like in the flask if it were possible for her to see it magnified many times. Then a pump is used to remove some of the air, and Jamie is asked to describe what the air would look like if half of the air were removed from the flask. The tasks examine whether Jamie spontaneously applies the idea of the particulate nature of air, whether she thinks that the particles are evenly scattered, and what she thinks is in the spaces between the particles. Jamie's responses are repeatedly prompted by the interviewer. In another interview Chris, a 6th-grader, explains what happens to the air in a syringe after it gets compressed. Chris is asked to draw a diagram of the syringe and the air inside before and after it gets squashed. The task examines whether Chris spontaneously applies the idea of the particulate nature of air, whether he thinks that the particles are evenly scattered, and what he thinks is in the spaces between the particles. These interviews and others provide a compelling exhibit of student ideas about the structure of matter.

The following example drawn from *Benchmarks*, Chapter 15: The Research Base, describes what cognitive scientists know—and do not know—about how students think about the nature of science in general and, more specifically, about how scientific knowledge is developed over time. This research is relevant to the science literacy goals presented in section 1A, The Scientific World View of *SFAA's* Chapter 1: The Nature of Science.

Research on students' understanding of the nature of science has been conducted for more than 30 years. The earlier part of the research investigated students' understanding about scientists and the scientific enterprise and about the general methods and aims of science (Cooley & Klopfer, 1961; Klopfer & Cooley, 1963; Mackey, 1971; Mead & Metraux, 1957; Welch & Pella, 1967). More recent studies have added students' understanding of the notion of "experimentation," the development of students' experimentation skills, students' understanding of the notions of "theory" and "evidence," and their conceptions of the nature of knowledge. The available research is reviewed in Lederman (1992).

Research on the nature of science focuses mainly on the middle-school and high-school grades. There are few studies that investigate what elementary-school learning experiences are effective for developing an understanding of the nature of science, although Susan Carey's and Joan Solomon's work is a beginning in that direction (Carey, Evans, Honda, Jay, & Unger, 1989; Solomon, Duveen, Scot, McCarthy, 1992).

Research in the 1960s and 70s used multiple-choice questionnaires. Recent studies using clinical interviews reveal discrepancies between researchers' and students' understanding of the questions and the proposed answers in those questionnaires. This finding raises doubt about the earlier studies' findings because almost none of them used the clinical interview to corroborate the questionnaires. Therefore, the following remarks draw mainly upon the results of the relatively recent interview studies.

1A The Scientific World View

Although most students believe that scientific knowledge changes, they typically think changes occur mainly in facts and mostly through the invention of improved technology for observation and measurement. They do not recognize that changed theories sometimes suggest new observations or reinterpretation of previous observations (Aikenhead, 1987; Lederman & O'Malley, 1990;

Waterman, 1983). Some research indicates that it is difficult for middle-school students to undestand the development of scientific knowledge through the interaction of theory and observation (Carey et al., 1989), but the lack of long-term teaching interventions to investigate this issue makes it difficult to conclude that students can or cannot gain that understanding at this grade level.

Aikenhead, G.S. (1987). High school graduates' beliefs about science-technology-society III. Characteristics and limitations of scientific knowledge. *Science Education*, 71, 459-487.

Lederman, N., & O'Malley, M. (1990). Students' perceptions of the tentativeness in science: Development, use, and sources of change. *Science Education*, 74, 225-239.

Waterman, M. (1983). Alternative conceptions of the tentative nature of scientific knowledge. In J. Novak (Ed.), *Proceedings of the international seminar misconceptions in science and mathematics* (pp. 282-291). Ithaca, NY: Cornell University.

Carey, S., Evans, R., Honda, M., Jay, E., & Unger, C. (1989). An experiment is when you try it and see if it works: A study of grade 7 students' understanding of the construction of scientific knowledge. *International Journal of Science Education*, 11, 514-529.

Cooley, W., & Klopfer, L. (1961). *Test on understanding science*, Form W. Princeton: Educational Testing Service.

Klopfer, L., & Cooley, W. (1963). Effectiveness of the history of science cases for high schools in the development of student understanding of science and scientists. *Journal of Research in Science Teaching*, 1, 35-47.

Mackey, L. (1971). Development of understanding about the nature of science. *Journal of Research in Science Teaching*, 8, 57-66.

Mead, M., & Metraux, R. (1957). Image of the scientist among high-school students: A pilot study. *Science*, 26, 384-390.

Welch, W., & Pella, M. (1967). The development of an instrument for inventorying knowledge of the processes of science. *Journal of Research in Science Teaching*, 5, 64-68.

Lederman, N. (1992). Students' and teachers' conceptions of the nature of science: A review of the research. *Journal of Research in Science Teaching*, 29, 331-359.

Carey, S., Evans, R., Honda, M., Jay, E., & Unger, C. (1989). An experiment is when you try it and see if it works: A study of grade 7 students' understanding of the construction of scientific knowledge. *International Journal of Science Education*, 11, 514-529.

Solomon, J., Duveen, J., Scot, L., & McCarthy, S. (1992). Teaching about the nature of science through history: Action research in the classroom. *Journal of Research in Science Teaching*, 29, 409-421.

NOTES ON THE COGNITIVE RESEARCH DATABASE ON CD-ROM

Resources in the research database can be accessed through an alphabetical listing of articles, books, and videos by author and title or by selecting research related to a specific *SFAA* chapter or section. The references to cognitive research literature included in Chapter 15 of *Benchmarks for Science Literacy* also can be located alphabetically by author or by *SFAA* chapters.

FUTURE VERSIONS

As with the trade book and college courses databases, this research database is far from exhaustive. Rather, it provides examples of the kinds of resources that can help teachers better understand how their students think and learn about science. In searching for research articles written for teacher journals, Project 2061 staff had little trouble locating articles that address student understanding of mathematics and the physical sciences, but found far fewer articles related to student understanding of the life sciences, and fewer still on student understanding of concepts in social studies, technology, the nature and history of science, and scientific inquiry. For some topics, such as controlling variables, no research articles whatsoever could be found in teacher journals. Most, but not all such topics underrepresented in the teacher journals reflect existing gaps in the research base.

By publishing *Resources for Science Literacy: Professional Development* and alerting educators and publishers to the neglected topics, Project 2061 hopes to stimulate more good articles in teacher journals on research related to science literacy. Project 2061 welcomes suggestions for additional articles, books, or videos related to cognitive research issues and will consider them for future editions.

Send suggestions to:

Project 2061
American Association for the Advancement of Science
1333 H Street, NW
P.O. Box 34446
Washington, D.C.20005
FAX: 202/842-5196
Electronic Mail: project2061@aaas.org (please
   identify subject as "research")

ALSO SEE ☞ For more detailed instructions on using the cognitive research database, please refer to Chapter 7: Using the *Resources for Science Literacy* CD-ROM on page 107 of this volume.

JACOB ARMSTEAD LAWRENCE, *Builders No. 1, 1970.*

The college courses database included on the *Professional Development* CD-ROM provides 15 syllabi from college courses that were designed to increase students' knowledge of science, mathematics, and technology and their interconnections. The courses chosen were not necessarily designed with *Science for All Americans* in mind, but were submitted by university faculty who had analyzed their existing syllabi for strong links to *Science for All Americans*. As with the Science Trade Books component of the *Professional Development* CD-ROM, a special effort was made to include course syllabi that cover those topics found in *Science for All Americans* with which teachers are least likely to be familiar—technology, cross-cutting themes, and the nature and history of science, for example. Depth, rather than breadth, of content coverage was another consideration for inclusion in the database.

The syllabi presented on the *Professional Development* CD-ROM represent a small sampling of similar courses offered around the country and are intended to stimulate discussions about how to enhance the science literacy of prospective teachers. They might also serve as a guide for teachers who want to explore on their own a specific area of science, mathematics,

or technology. Project 2061 is eager to identify additional syllabi for future versions of *Resources for Science Literacy: Professional Development* and invites faculty to submit course syllabi that are based on science literacy goals.

A brief summary of each syllabus follows, along with a description of the eight sections that comprise the college courses database and some illustrative examples selected from several of the syllabi.

ABOUT THE SYLLABI

To identify course syllabi for inclusion in the database, Project 2061 drew on three sources:

- The list of courses developed or adapted for The New Liberal Arts Program (funded by the Sloan Foundation in the 1980s) to integrate ideas from the natural and social sciences, technology, humanities, and the arts.[1]

- *The Liberal Art of Science: Agenda for Action* (AAAS, 1990), which contains recommendations on how to make science part of everyone's undergraduate liberal education.[2]

---

[1] Extended syllabi from the New Liberal Arts Program are available from the Department of Technology and Society, State University of New York, Stony Brook, NY 11794-2250.

[2] American Association for the Advancement of Science. (1990). *The Liberal Art of Science: Agenda for Action. Washington, D.C.:* Author.

- Responses from university faculty who were already using *Science for All Americans* as the basis for their programs and were invited to submit detailed descriptions of courses designed to increase students' understanding of science, mathematics, technology, and their interconnections.

COURSE SUMMARIES

The following brief summaries of the syllabi indicate the wide range of approaches to the science, mathematics, and technology content represented in the database. They also describe how the courses specifically address ideas from *Science for All Americans (SFAA)*.

Learning Science as Inquiry: The Biology and Chemistry of Fat

Merle Bruno and Nancy Lowry, Hampshire College
One of a series within the Hampshire College program *Learning Science as Inquiry*, this first-year course is designed for students of all backgrounds. The program is intended to attract students to science through concrete experiences on topics of interest to them, to develop their skills in analysis and quantitative reasoning, and to introduce them to how scientists ask questions and view the nature of science. In *The Biology and Chemistry of Fat*, students see what biologists and chemists have to say about fat in and out of the body and study some fats in the lab. The class reads and discusses primary and secondary literature from a booklet prepared for the course. Students choose their favorite fat questions to work on and present their findings to the class and in papers. The course addresses, among others, topics from *SFAA* Chapter 1: The Nature of Science, Chapter 6: The Human Organism, and Chapter 12: Habits of Mind.

Engineering—Intermediate Technology

Chris Bull and Barrett Hazeltine, Brown University
This course examines the applications of technology, particularly small-scale approaches, to real-world problems and evaluates alternative approaches in terms of social impact, influence on self-reliance, cost, and environmental effects. It is elective for all students, but is intended primarily for nonscience majors. The course seeks to foster an understanding of the interactions of technology, science, and society by focusing on a particular technological approach. Another purpose is to help students feel comfortable when evaluating issues with a major technological component. Students acquire a better understanding of concepts and the nature of technology by designing and implementing projects intended to solve specific problems or meet particular needs. The course focuses on topics from *SFAA* Chapter 3: The Nature of Technology and Chapter 8: The Designed World.

Evolution—Zoology/Botany

Mark Hafner, Louisiana State University, and Sherry Southerland, University of Utah
This is a two-course series—a lecture and an associated laboratory course—on the basic principles and processes of evolutionary biology designed for stu-

dents majoring in the natural sciences. The course's goal is for students to understand not only the basic concepts that comprise the theory of evolution, but also the linkages between these concepts. Lectures use animal and plant examples to illustrate and clarify fundamental concepts in evolution. Both classical and more recent contributions to our understanding of evolution are analyzed. Students reinforce their understanding of the concepts introduced in lectures by applying them to biological situations in the laboratories. The laboratories include simulations, exercises, demonstrations, and discussions that emphasize the historical nature of evolutionary biology and address students' conceptual difficulties documented in the cognitive research literature. The course focuses on topics from *SFAA* Chapter 1: The Nature of Science, Chapter 5: The Living Environment, and Chapter 10: Historical Perspectives. Topics related to evolution and natural selection are addressed at a higher level of sophistication than the fundamental literacy description in *SFAA*.

Nature of Science and Its Interactions with Technology/Society

Norman Lederman, Oregon State University
This course is designed for secondary level in-service teachers and science education doctoral students to develop their understanding of the history, nature, and philosophy of science, as well as the interactions among science, technology, and society. A second purpose is to enable students to apply such knowledge to the development of instructional materials for use in secondary schools. Students clarify their views on the nature of science through an integrated series of readings, discussions, demonstrations, and hands-on activities. The course addresses several topics from *SFAA* Chapter 1: The Nature of Science.

Physics by Inquiry

Lillian McDermott, University of Washington
Physics by Inquiry is a set of laboratory-based modules that provide a step-by-step introduction to physics and the physical sciences. It has been developed through an iterative, interactive process of research, curriculum development, and instruction. The modules have been developed and class-tested mainly in special courses to prepare precollege teachers to teach science as a process of inquiry. The modules have also been used to help underprepared students succeed in the mainstream science courses that are a gateway to science-related careers. Through in-depth study of simple physical systems and their interactions, students gain direct experience with the process of science. Starting from their own observations, students develop basic physical concepts, use and interpret different forms of scientific representa-

For a description of the topics covered in each chapter of *SFAA*, see this volume's Chapter 1: About *Science for All Americans*, beginning on page 3.

ALSO SEE

tions, and construct explanatory models with predictive capability. All of the modules have been explicitly designed to develop scientific reasoning skills and to provide practice in relating scientific concepts, representations, and models to real-world phenomena. The modules address, among others, topics from *SFAA* Chapter 1: The Nature of Science, Chapter 4: The Physical Setting, Chapter 11: Common Themes, and Chapter 12: Habits of Mind.

The Theory and Practice of Science—Biology
Robert Pollack, Columbia University

This is a course designed for nonscience majors fulfilling a portion of their science distribution requirement. By rigorously examining particular developments and discoveries of contemporary biology, the course acquaints the nonscientist with what scientists do; how scientists approach technical, methodological, and philosophical problems; and how the theory and practice of science evolve together. The course introduces sufficient mathematics so that original papers may be used as a basis for study. The course addresses, among others, topics from *SFAA* Chapter 1: The Nature of Science, Chapter 5: The Living Environment, and Chapter 10: Historical Perspectives.

Foundations of Science
Ezra Shahn, Hunter College of the City University of New York

Designed as an introduction to science for nonscience majors, this one-year multidisciplinary science course focuses on a limited number of concepts that lie at the foundations of science. The three major themes considered are: the Heliocentric Theory and the Study of Motion, the Nature and Properties of Matter, and the History of Earth and of Life. Although it is not a history of science course, themes are developed from a predominantly historical point of view. Integral to the course are laboratory periods which are intended to enhance the development of cognitive skills and an understanding of the nature of inquiry and the scientific enterprise. A "learning cycle" approach to structuring lab experiences is used in which the student explores, analyzes, and applies knowledge. The course addresses, among others, topics from *SFAA* Chapter 1: The Nature of Science, Chapter 4: The Physical Setting, Chapter 5: The Living Environment, and Chapter 10: Historical Perspectives.

Science/Technology/Society Interaction
Barbara Spector, University of South Florida

This course is designed to develop students' awareness of science and technology as human enterprises that take place in a social, environmental, and historical context. Various interactions of science, technology, and society are explored in the context of issues relevant to the students. The goal of the course is to enable students to construct a historical and philosophical understanding of (a) the nature of the scientific enterprise, including the interaction of science, technology, and society; (b) the multiple dimensions and complexities of sample science/technology/society (STS) topics; and (c) how to teach STS to diverse audiences. The course is required of all science education majors. It addresses, among others, topics in *SFAA* Chapter 1: The Nature of Science and Chapter 3: The Nature of Technology.

Scientific and Technological Literacy Program: Matter, Energy, Life, and Systems

Victor Stanionis, Iona College

The first part of a two-course program in science and technology, *Matter, Energy, Life, and Systems* is required of all freshmen at Iona College. It is designed to develop the foundation for science literacy necessary to deal with technology-related problems in a modern society. Basic scientific concepts underlying matter, energy, and life are introduced, and students are helped to develop the reasoning and problem-solving skills characteristic of scientists. After completing this introductory course, students choose a second course from five options: commercial systems, environment, health, energy, and computer music. *Matter, Energy, Life, and Systems* provides a survey of topics from *SFAA* Chapter 1: The Nature of Science, Chapter 2: The Nature of Mathematics, Chapter 3: The Nature of Technology, Chapter 4: The Physical Setting, Chapter 5: The Living Environment, Chapter 6: the Human Organism, Chapter 8: The Designed World, Chapter 11: Common Themes, and Chapter 12: Habits of Mind.

The following five courses are options for the second half of Iona College's required two-course program in science and technology.

Scientific and Technological Literacy Program: Commercial Systems Theme

Victor Stanionis, Iona College

This course emphasizes the roles of scientific analysis, scientific knowledge, and technology as vital tools in managerial decision making in business. Business-related case studies are used to demonstrate that science and scientific methods are components of typical business decision-making problems. Students have the opportunity to apply some of their basic skills and emerging literacy to interesting problems that illustrate many aspects of science and technology at work in society. The course addresses, among others, topics from *SFAA* Chapter 3: The Nature of Technology, Chapter 8: The Designed World, Chapter 9: The Mathematical World, Chapter 11: Common Themes, and Chapter 12: Habits of Mind.

Scientific and Technological Literacy Program: Environmental Theme

Victor Stanionis, Iona College

The focus of this course is the study of science and technology in the context of societal concerns about problems related to waste, matter, and energy. A variety of waste types produced by modern society are studied with emphasis on management techniques, toxic effects, and recycling possibilities. The course employs the student's understanding of science and technology and of some environmental problem areas to appraise alternative futures. Technology assessment and systematic forecasting methods are studied. Case studies and/or individual student projects are used. The course addresses, among others, topics from *SFAA* Chapter 1: The Nature of Science, Chapter 3: The Nature of Technology, Chapter 8: The Designed World, Chapter 9: The Mathematical World, Chapter 11: Common Themes, and Chapter 12: Habits of Mind.

Scientific and Technological Literacy Program: Health Theme

Victor Stanionis, Iona College

In this course the emphasis is on understanding the interactive effects of biological, psychological, social and environmental factors on mental health and illness. The course gives special consideration to practical and social issues involving the development and enhancement of mental health and employs the student's understanding of science and technology and some current health problem areas in order to appraise alternative futures. Technology assessment and systematic forecasting methods are studied. Case studies and/or individual student projects are used. The course addresses, among others, topics from *SFAA* Chapter 3: The Nature of Technology, Chapter 6: The Human Organism, Chapter 8: The Designed World, Chapter 9: The Mathematical World, Chapter 11: Common Themes, and Chapter 12: Habits of Mind.

Scientific and Technological Literacy Program: Energy Theme

Victor Stanionis, Iona College

Societal concerns about problems related to the production, distribution, and uses of electrical energy provide the context for the study of science and technology. The flow of electricity in an industrial society is examined with an emphasis on the fuels, production processes, efficiency, impact, and safety of this technology. The course employs the student's understanding of science, technology and some current energy problem areas in order to appraise alternate futures. Technology assessment and systematic forecasting methods are studied. Case studies and/or individual student projects are used. The course addresses, among others, topics from *SFAA* Chapter 3: The Nature of Technology, Chapter 4: The Physical Setting, Chapter 8: The Designed World, Chapter 9: The Mathematical World, Chapter 11: Common Themes, and Chapter 12: Habits of Mind.

Scientific and Technological Literacy Program: Computer Music Systems

Victor Stanionis, Iona College

Using a MIDI (Musical Instrument Digital Interface) system, this course is designed to develop the foundation for science literacy through the study of computer music. The course employs the student's understanding of science, technology, and systems to assess the societal impact of computer music along with its costs, benefits, and detriments. The course addresses, among others, topics from *SFAA* Chapter 2: The Nature of Mathematics, Chapter 3: The Nature of Technology, Chapter 4: The Physical Setting, Chapter 8: The Designed World, Chapter 9: The Mathematical World, Chapter 11: Common Themes, and Chapter 12: Habits of Mind.

The Sciences: Approaches to the Natural World

Jerold Touger, Curry College

This course is an interdisciplinary (biology, chemistry, physics, astronomy) introduction to science that examines how scientists formulate and address questions about life, matter, and the nature of the universe. It is intended for all students except science and nursing majors. The course encourages students to consider: (1) elements of critical thinking; (2) the nature of scientific thinking and "doing"; (3) the interrelatedness of the sciences; (4) the utility of science; and (5) sociopolitical implications of science and technology. Emphasis is on methods by which scientists achieve understanding and perspectives on the contemporary world that such understanding provides. The course addresses, among others, topics from *SFAA* Chapter 1: The Nature of Science, Chapter 3: The Nature of Technology, Chapter 4: The Physical Setting, Chapter 5: The Living Environment, Chapter 6: The Human Organism, Chapter 8: The Designed World, Chapter 9: The Mathematical World, Chapter 11: Common Themes, and Chapter 12: Habits of Mind.

Framework for Analyzing Syllabi

In developing the database of college courses and providing some guidelines for contributors, Project 2061 confronted the question of what it means for a course to address the ideas in *SFAA*. The first consideration was the extent of the congruence between topics in the course and in *SFAA*. The guidelines also attempted to provide contributors with a basis for estimating the likelihood that students would learn the concepts and ideas that are targeted in the course.

From these considerations, the following series of questions emerged as a framework for analyzing or creating courses that are centered on *SFAA* concepts:

- Is each topic or concept within a topic treated in adequate depth to realistically expect students to learn it?
- Is attention called to appropriate connections among concepts presented in the course? In other courses?
- Does the course attempt to address student misconceptions that have been summarized in *Benchmarks for Science Literacy* Chapter 15: The Research Base?

- Is the course structured to make possible the collection of evidence on student understanding of the concepts targeted? Is the evidence used to revise the course?
- Are students engaged in activities that give them first-hand experiences with concepts and then given opportunities to reflect on their activities?
- Are opportunities provided for students to apply their knowledge in varied contexts—e.g., explaining everyday phenomena or considering alternative solutions to practical problems?

ORGANIZATION OF THE COLLEGE COURSES DATABASE

Each syllabus submitted for *Resources for Science Literacy: Professional Development* was reorganized into eight sections:

1. a general description of the course;
2. information about the author or contact person;
3. resources for the course;
4. links between the course and *SFAA*;
5. a detailed explanation of those links;
6. an extended syllabus, if available;
7. a short summary of how the course design supports student learning; and
8. any research related to student understanding of the course content.

At a minimum, each syllabus includes information on the first four sections, but many authors were also able to provide additional information for the last four as well. Though some syllabus authors chose not

to submit lengthy analyses for the database, several did describe their courses in terms of the analytical framework discussed earlier, and their responses can be accessed in the section that deals with student learning.

Examples from the CD-ROM

To give a sense of the kinds of material likely to be encountered in the college courses database, examples for each section (except section 6) have been selected from several different syllabi and are presented below. A vertical blue line in the margins indicates these excerpts from the syllabi.

1. A general description of the course—topics included, the sequence of those topics, and student assignments. The following example is drawn from *Engineering — Intermediate Technology*, a course developed by Bull and Hazeltine at Brown University.

Engineering—Intermediate Technology
Rationale: Different technological solutions can meet the same human need. One way to heat a house is a nuclear power plant supplying a resistive heater. Another way is a solar collector and ample insulation. These technologies can be compared on the basis of cost, self-dependence, encouragement of self-reliance, or environmental effects. Is a solar heated house desirable simply because the owner can understand it and build it herself/himself and because it does not pollute? This course considers whether a so-called "Intermediate Technology" on the one hand, meets relevant human physical needs

and, on the other hand, actually brings with it desirable personal and social benefits.

One part of the course consists of a discussion of possible benefits of different technological levels. Another part is a study of various intermediate technologies. The last part is an evaluation of these technologies. Do they work? Do they really promote a fulfilling, healthy life style? Do they really lead to a responsible and responsive government?

Schedule: The schedule below for the course is intentionally general, so its implementation can be adjusted to meet the wishes of participants.

Definition of Intermediate Technology and Rationale for Its Use	1.0 week
Water Power	1.0
Wind Power	0.5
Methane Digesters, etc.	0.5
Food Production/Consumption	1.0
Solar Energy	1.0
Human-Powered Vehicles	1.0
Other Intermediate Technologies	1.0
Technology in Colonial America and in Contemporary Southern Africa	1.0
Trips to Old Slater Mill, Southside Land Trust, etc.	1.0
Limitations of Technological Solutions	0.5
Guest Lecturers on Impact of Technology in Other Cultures	1.0
Implications of Intermediate Technology, especially in Contemporary U.S. Society	1.5
Presentation of Projects	1.0

Course requirements: The course requirements are participation in discussion, completion of problems—several from each chapter, as chosen—and a project in three parts. The first part consists of a proposal for the project—what will be done and why the concept is useful. The second part consists of a paper describing why the project makes sense— what the approach offers, particularly in terms of life style and the basic calculations and so forth. The third part consists of the project itself and a paper explaining it.

For example, if the project is small-scale gardening, the basic calculations would show the amount of land required, amount of food produced, cost and time for preparation and cultivation, cost of seeds and fertilizer, and so forth. The life-style section of the paper would discuss drawbacks of existing agriculture and advantages of what you are proposing. The project would be the plants you have grown and the plan for the garden.

Grades: The final grade weighs class participation and problems equally with the project. Projects are presented to the class during reading period. Some people feel shy about participating in discussion. If students prefer to write out their position on the topic being considered, those position statements can substitute for discussion.

2. Information about the author or contact person.

For more information about the course, please contact:

Dr. Barrett Hazeltine
Division of Engineering
Brown University
Providence, RI 02912
Phone: 401/863-2673
Fax: 401/863-1152
E-mail: hazeltin@brownvm.brown.edu

3. Resources for the course—reading lists, lab manuals, and any other published course materials. As an example, a portion of the reading list for Pollack's *The Theory and Practice of Science—Biology* follows:

Books

Darwin, C. (1968). *The origin of species by means of natural selection.* Harmondsworth: Penguin.

Pollack, R. (1994). *Signs of Life: The language and meanings of DNA.* Boston: Houghton Mifflin.

Schrodinger, E. (1992). *What is life? The physical aspect of the living cell.* New York: Cambridge University Press.

Watson, J. (1980). *The double helix: A personal account of the discovery of the structure of DNA.* New York: Norton.

Papers

Avery, O.T., Colin, M.M., & McCarty, M. (1944). Studies on the chemical nature of the substance inducing transformation of pneumococcal types. *Journal of Experimental Medicine,* 79, 137-158.

Ayala, F.J. (1992). Evolution. In B. Davis (Ed.), *The genetic revolution* (pp. 179-195). Baltimore: Johns Hopkins University Press.

Bhattacharyya, M.K., Smith, A.M., Ellis, T.H.N., Hedley, C., & Martin, C. (1990). The wrinkled-seed character of pea described by Mendel is caused by a transposon-like insertion in a gene encoding starch-branching enzyme. *Cell,* 60, 115-122.

Gordon, J.W., & Ruddle, F.H. (1981). Integration and stable germ line transmission of genes injected into mouse pronuclei, *Science,* 214, 1244-1246.

4. Links, identified by the course author, between the course and chapters of *Science for All Americans* and, in most cases, sections within chapters of *Science for All Americans.* For their course *Engineering—Intermediate Technology* (see summary above), Bull and Hazeltine of Brown University suggest the following links between syllabus topics (in italic below) and sections and topics within *SFAA* Chapter 3: The Nature of Technology (in boldface below):

3A: Technology and Science
  Technology Draws on Science and
    Contributes to It
    Advantages of Understanding
3B: Design and Systems
  The Essence of Engineering Is Design
    Under Constraint
    Designs, Constraints, and Trade-offs
    Issues of Use and Maintenance
  All Technologies Involve Control
    Human Interfaces
3C: Issues in Technology
  The Human Presence
    Benefits and Side Effects of Technology
  Technological and Social Systems Interact
    Strongly
  The Social System Imposes Some Restrictions
    on Openness in Technology
    Commercial Control of Innovations
  Decisions about the Use of Technology
    Are Complex
    Decisions about Technology
    Alternatives, Responsibility, and Uncertainty

5. A detailed explanation of how the course links to SFAA. In the following excerpt from the syllabus for *Evolution—Zoology/Botany*, Hafner and Southerland describe how their course links to *SFAA* Chapter 5: The Living Environment. They also reflect on some of the interconnections among topics and evaluate the level of difficulty of their course's content:

Within the *SFAA* chapter dedicated to the living environment, it is not surprising that the majority of the course addressed the section on the evolution of life. However, topics within two related sections, Diversity of Life and Heredity, also formed major components of the course. Classification, a topic in the section Diversity of Life, was a central focus of the class. Instruction emphasized species concepts and the related assumptions that form the basis of each of the various research programs. While the treatment of species concepts was an excellent avenue to display the tentative and self-correcting nature of science, this topic also provided students with a deep understanding of the dynamics of natural populations and lead them to a later understanding of the process of speciation. As was the case with much of the content of the class, the treatment of species concepts far exceeded the complexity reflected in the *SFAA* document.

6. An extended syllabus. In most cases, an extended syllabus is available from the course developers or from their institutions.

7. A short summary of how the course design supports student learning. In this excerpt from *Evolution—Zoology/Botany*, Hafner and Southerland describe how their course addresses student misconceptions related to evolution. Their discussion draws on *Benchmarks* Chapter 15: The Research Base:

Many of these strongly held alternative conceptions were addressed through attention to the historical development of evolutionary theory (as proposed by Jensen & Finely, 1993), others were addressed through laboratory activities (such as the bead-bug activity described by Bishop & Anderson, 1985), and others were addressed through an explanation of the scientific conception compared to the alternative explanation.

The first two pedagogical practices, the historical development and laboratory activities, have been shown to be reasonably effective in research studies (Jensen & Finley, 1993; Zuzovsky, 1994; Bishop & Anderson, 1990). The third option, a rationalistic analysis of the scientific conception, is the traditional approach, which has been shown to be less effective in displacing strongly held alternative conceptions (Bishop & Anderson, 1990; Demastes et al., 1992; Demastes, Good, & Settlage, 1995).

8. Any research (published or not) related to student understanding of the course content. The following excerpt from McDermott's *Physics by Inquiry* includes a few citations from the course's related bibliography:

Physics by Inquiry is the product of an intensive, collaborative effort by the Physics Education Group in the Physics Department at the University of Washington. The group includes faculty, research associates, and graduate students. Members of the group conduct in-depth investigations of student understanding through which they

identify and analyze specific difficulties that students encounter in studying physics. This research has provided the foundation for the design of the instructional strategies that are incorporated in *Physics by Inquiry*.

McDermott, L.C. (1991). What we teach and what is learned: Closing the gap. *Am. J. Phys., 59* (4) 301 (1991).

McDermott, L.C., Piternick, L., & Rosenquist, M. (1980). Helping minority students succeed in science: I. Development of a curriculum in physics and biology; II. Implementation of a curriculum in physics and biology; III. Requirements for the operation of an academic program in physics and biology. *J. Coll. Sci. Teach.*, 9, 135, 201, 261.

McDermott, L.C., Rosenquist, M.L., & van Zee, E.H. (1987). Student difficulties in connecting graphs and physics: Examples from kinematics. *Am. J. Phys., 55* (6), 503.

McDermott, L.C., & Shaffer, P. (1992). Research as a guide for curriculum development: An example from introductory electricity, Part I: Investigation of student understanding. *Am. J. Phys., 60* (11), 994; Erratum to Part I, *Am. J. Phys., 61* (1), 81.

Rosenquist, M.L., & McDermott, L. (1987). A conceptual approach to teaching kinematics. *Am. J. Phys., 55* (S), 407.

Shaffer, P., & McDermott, L. (1992). Research as a guide for curriculum development: An example from introductory electricity, Part II: Design of instructional strategies. *Am. J. Phys., 60* (11), 1003.

Notes on the College Courses Database

Though each syllabus is organized into the same eight sections mentioned above, the 15 submissions differ greatly in emphasis, style, and level of detail. Rather than forcing the syllabi into a fixed standard style, some irregularities remain, and these require some explanation:

Searching. Courses in the database can be accessed by title. Each section of the course—e.g., syllabus, bibliography, and links to *SFAA*—can then be accessed individually.

Links to *SFAA*. Some authors have specified chapters or sections from *SFAA,* and some have specified sub-section topics, indicating where in the course these topics are addressed. Others indicated all the topics from *SFAA* that their course in some way covers, without making explicit how the course does so. Still others supplied a narrative describing how particular activities within their course address *SFAA* sections or topics.

Instructional strategies. Some of the syllabus authors explained how their courses incorporate appropriate instructional strategies to help students make progress toward these goals. Three authors provided references to research related to student understanding of course content.

Future Versions

For the next edition of the college courses database, contributors will be encouraged to provide more information. For example, they will be asked to identify explicitly where their course addresses specific topics from *SFAA* and to address pedagogical issues in more detail. The next edition will also include many more syllabi.

Project 2061 will encourage arts and science and engineering departments and others responsible for the content knowledge of teachers to use the analytical framework described above to design and modify courses around *SFAA* (or other sets of learning goals) and to submit the course syllabi. In addition, the next version will include syllabi from education courses designed to prepare teachers to better understand student learning through use of *Benchmarks for Science Literacy*.

Send suggestions to:

Project 2061
American Association for the Advancement of Science
1333 H Street, NW
P.O. Box 34446
Washington, D.C. 20005
FAX: 202/842-5196
Electronic Mail: project2061@aaas.org (please
   identify subject as "college courses")

For more detailed instructions on using the college courses database, please refer to Chapter 7: Using the *Resources for Science Literacy* CD-ROM on page 107 of this volume.

ALSO SEE

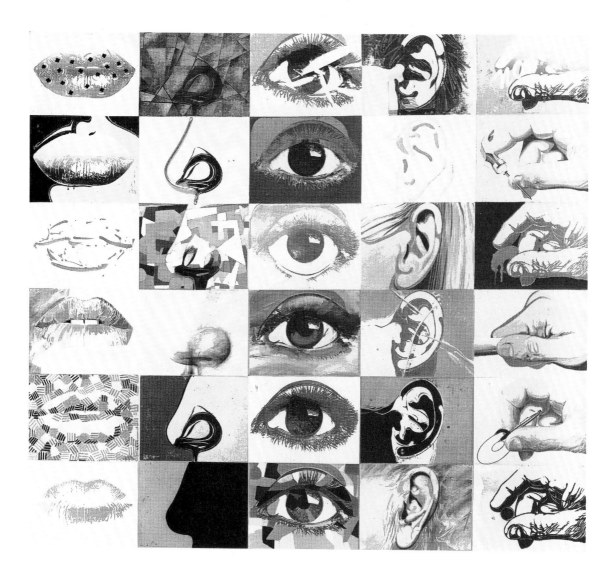

PAUL GIOVANOPOULOS, *Five Senses, 1990.*

Chapter 5 COMPARISONS OF *BENCHMARKS* TO NATIONAL STANDARDS

Shortly after Project 2061 published *Science for All Americans* in 1989, the National Council for Teachers of Mathematics released its *Curriculum and Evaluation Standards for School Mathematics*. As Project 2061 worked to translate *Science for All Americans'* science literacy goals into learning goals or "benchmarks" for the ends of grades 2, 5, 8, and 12, the National Research Council (NRC) of the National Academy of Sciences, and the National Council for the Social Studies (NCSS) were engaged in similar efforts to prepare national content standards. In 1993 Project 2061 published *Benchmarks for Science Literacy*. The NCSS published its *Curriculum Standards for Social Studies* in 1994, and the NRC published its *National Science Education Standards (NSES)* for grades K-12 at the end of 1995.

Developed by experts in science, mathematics, and technology as well as teachers, these documents share some important characteristics: (1) they provide an explicit set of K-12 learning goals; (2) they recommend goals that are developmentally appropriate for students; and (3) they provide guidance in reducing curriculum content so that the most important ideas can be learned well by all. The documents differ in focus, organization, level of detail, grade-levels for learning goals, targeted audience, and, to some degree, in strategy for reform.

Project 2061 has analyzed the three standards documents and made a detailed comparison of each with *Science for All Americans* and *Benchmarks for Science Literacy*. These three comparisons are included on the *Resources for Science Literacy: Professional Development* CD-ROM so that educators can make more informed use of the documents together or separately. Educators will want to use the comparisons in a variety of ways—for example, to evaluate curriculum frameworks, materials, and instruction for their match to benchmarks or standards.

About the Comparisons

All three comparisons reveal the considerable overlap between *Benchmarks* and standards; this should reassure teachers that experts in science, mathematics, and social studies *agree* broadly on what concepts and skills are important for all students to learn. Most school districts and state curriculum committees do not have the time or the resources that the AAAS, NRC, NCTM, and NCSS had to develop sound goal statements consistent with expert views of these fields and the latest research about what children of various ages can learn. District and state curriculum planners may want to begin with the ideas on which *Benchmarks* and the national standards are in accord. The areas where the documents differ can serve as the basis for discussion about how local students will best be served.

The following descriptions, excerpts, and examples provide a brief introduction to the three comparisons presented on the *Professional Development* CD-ROM.

COMPARISON OF *BENCHMARKS* AND *NATIONAL SCIENCE EDUCATION STANDARDS*

This detailed analysis of how Project 2061's *Benchmarks* relates to the National Research Council's *National Science Education Standards (NSES)* offers comparisons at two levels: an organizational and topical comparison between the twelve *Benchmarks* chapters and the eight *NSES* content standard categories (Figure 1) and a substantive comparison between individual benchmarks and the *NSES* "fundamental concepts" and "fundamental abilities." The comparison of *Benchmarks* to national standards presented on the *Professional Development* CD-ROM includes

- a summary report describing how the comparison between *Benchmarks* and *NSES* was made and identifying the most important findings,
- a set of tables and charts illustrating how *Benchmarks* and *NSES* are—and are not—congruent,

- a listing of benchmarks linked to the relevant fundamental concepts and fundamental abilities found in *NSES*, and
- a listing of *NSES*' fundamental concepts and fundamental abilities linked to the relevant benchmarks.

Starting with either set of learning goals, users can explore both *Benchmarks* and *NSES* and can easily shift from one to the other. The following highlights the findings of the analysis and presents some excerpts from the comparison of individual benchmarks to fundamental concepts and abilities.

A Consensus on Content

Initially prepared by Project 2061 staff as background for its solicited group review of the *NSES* draft, the comparative analysis reveals that *Benchmarks* and *NSES* are very similar in philosophies, language, difficulty and grade placement of their learning goals, and more. For example:

- Both documents share a commitment to reducing the sheer number of topics students must know to allow time for them to learn the most important ideas.
- In offering a common core of ideas and understandings about science and technology for all students, both exclude from basic literacy a host of topics—Ohm's law, series and parallel circuits, phyla of plants and animals, cloud types, balancing chemical reactions, and many others—that clutter the traditional science curriculum.
- Both documents emphasize understanding ideas over memorizing technical terms.

Figure 1: Correspondence of Content Divisions

BENCHMARKS CHAPTERS

BENCHMARKS Chapters	Unifying Concepts and Processes	Science as Inquiry	Physical Science	Life Science	Earth and Space Science	Science and Technology	Science in Personal and Social Perspectives	History and Nature of Science
Nature of Science		●				●	●	●
Nature of Mathematics		●						
Nature of Technology		●				●	●	
Physical Setting			●		●		●	
Living Environment				●			●	
Human Organism				●			●	
Human Society							●	
Designed World						●	●	
Mathematical World								
Historical Perspectives						●		●
Common Themes	●							
Habits of Mind		●				●		

NSES CONTENT STANDARDS: Unifying Concepts and Processes · Science as Inquiry · Physical Science · Life Science · Earth and Space Science · Science and Technology · Science in Personal and Social Perspectives · History and Nature of Science

• Both describe most ideas and topics at the same levels of difficulty and detail, placing them within comparable grade ranges.

The agreement between the documents is due in part to the close working relationship between Project 2061 and the National Research Council (NRC). Project 2061 shared drafts of *Benchmarks* with the NRC curriculum working groups as they developed the standards, and several Project 2061 staff members served on committees responsible for overseeing the development of the *NSES*. In the introduction to the *NSES*, the NRC acknowledges this special relationship:

"The many individuals who developed the content standards sections of the *National Science Education Standards* made independent use and interpretation of the statements of what all students should know and be able to do that are published in *Science for All Americans* and *Benchmarks for Science Literacy*. The National Research Council of the National Academy of Sciences gratefully acknowledges its indebtedness to the seminal work by the American Association for the Advancement of Science's Project 2061 and believes that use of *Benchmarks for Science Literacy* by state framework committees, school and school-district curriculum committees, and developers of instructional and assessment materials complies fully with the spirit of the content standards." (National Research Council, 1996)

Together, *Benchmarks* and *NSES* represent a strong national consensus among educators and scientists on what all K-12 students need to know and be able to do in science, mathematics, and technology. They provide states and school districts with a solid conceptual basis for reforming K-12 science education.

Differences between *Benchmarks* and *NSES*

Despite the high level of agreement between *Benchmarks* and *NSES*, the comparative analysis reveals some differences too. With regard to organization, for example, the two documents vary. *Benchmarks* organizes its four grade-level bands, K-2, 3-5, 6-8, and 9-12, into topic areas, whereas the *NSES* organizes its topic areas into three grade-level bands, K-4, 5-8, and 9-12. The territory they cover also differs (Figure 2). *Benchmarks* focuses exclusively on content and includes learning goals for both natural and social sciences and also for mathematics and technology. The *NSES*, on the other hand, provides content standards for natural science only, but also includes standards for teaching, professional development, assessment, programs, and systems. *Benchmarks* and *NSES* do not always cluster or separate ideas similarly (Figure 3), nor do they always place ideas at the same grade level.

Another difference between *NSES* and *Benchmarks* is in their treatment of scientific inquiry. Both agree that students should understand some important things about scientific investigation and both agree that there are some skills (e.g., keeping decipherable

Figure 2: Common Ground: Project 2061 and *NSES*

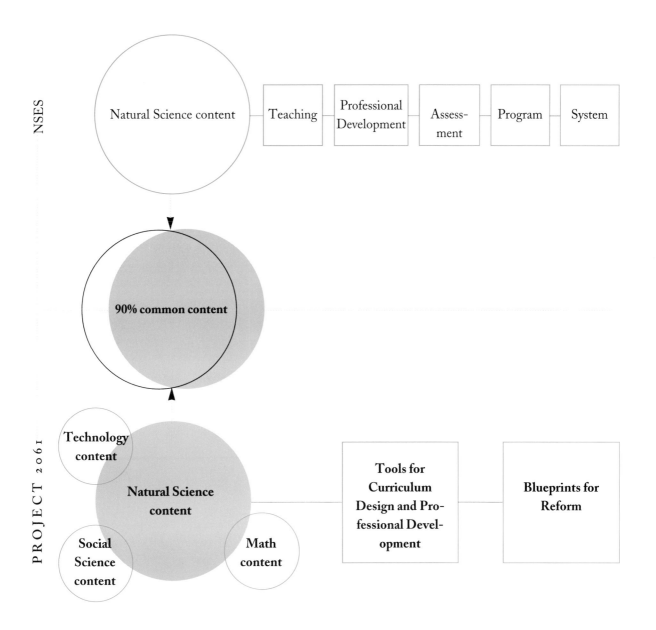

Figure 3: Primary Correspondence of a Single *NSES* Topic to *Benchmarks* Sections

BENCHMARKS: THE LIVING ENVIRONMENT

Diversity of Life	
Heredity	●
Cells	●
Interdependence of Life	
Flow of Matter and Energy	●
Evolution of Life	●

BENCHMARKS: THE HUMAN ORGANISM

Human Identity	
Human Development	●
Basic Functions	●
Learning	
Physical Health	
Mental Health	

NSES: Life Science (9–12)

The Cell

notes, sketching comprehensible graphs, constructing a reasonable argument) related to inquiry that all students should retain beyond school as part of science literacy. *NSES*, however, goes further by specifying the ability to *perform* scientific inquiry as one of its content standards. In contrast, *Benchmarks* emphasizes scientific inquiry as part of a range of critical thinking skills and other habits of mind characteristic of a science-literate adult. The same kinds of differences in emphasis appear in the two documents' treatment of design, the history of science, and the nature of science.

Examples from the CD-ROM

The following excerpts (indicated by a vertical blue line in the margins) taken from the *Resources for Science Literacy: Professional Development* CD-ROM illustrate how the analysis matches individual learning goals in *Benchmarks* to relevant ones found in *NSES*. Some of the differences described earlier (e.g., in aggregating or separating ideas or in grade placement) as well as the similarities, between the two sets of learning goals should also be apparent. Examples have been chosen to represent different topics and grade range. Learning goals from *Benchmarks* are shown first, followed by those from *NSES*.

Benchmarks **5F The Living Environment: Evolution of Life**
Grades K-2, page 123
Different plants and animals have external features that help them thrive in different kinds of places.
NSES Content Standard C Life Science: The Characteristics of Organisms
Grades K-4, page 129
Organisms have basic needs. For example, animals need air, water, and food; plants require air, water, nutrients, and light. Organisms can survive only in environments in which their needs can be met. The world has many different environments, and distinct environments support the life of different types of organisms.

Benchmarks **5E The Living Environment: Flow of Matter and Energy**
Grades K-2, page 119
Plants and animals both need to take in water, and animals need to take in food. In addition, plants need light.
NSES Content Standard C Life Science: The Characteristics of Organisms
Grades K-4, page 129
Organisms have basic needs. For example, animals need air, water, and food; plants require air, water, nutrients, and light. Organisms can survive only in environments in which their needs can be met. The world has many different environments, and distinct environments support the life of different types of organisms.

Benchmarks **5D The Living Environment: Interdependence of Life**
Grades 3-5, page 116
For any particular environment, some kinds of plants and animals survive well, some survive less well, and some cannot survive at all.
NSES Content Standard C Life Science: The Characteristics of Organisms
Grades K-4, page 129
Organisms have basic needs. For example, animals need air, water, and food; plants require air, water, nutrients, and light. Organisms can survive only in environment in which their needs can be met. The world has many different environments, and distinct environments support the life of different types of organisms.
NSES Content Standard C Life Science: Organisms and their Environments
Grades K-4, page 129
An organism's patterns of behavior are related to the nature of that organism's environment, including the kinds and numbers of other organisms present, the availability of food and resources, and the physical characteristics of the environment. When the environment changes, some plants and animals survive and reproduce, and others die or move to new locations.

Benchmarks 4F The Physical Setting: Motion
Grades 3-5, page 89
Changes in speed or direction of motion are caused by forces. The greater the force is, the greater the change in motion will be. The more massive an object is, the less effect a given force will have.

NSES Content Standard B Physical Science: Position and Motion of Objects
Grades K-4, page 127
The position and motion of objects can be changed by pushing or pulling. The size of the change is related to the strength of the push or pull.

Benchmarks 6E The Human Organism: Physical Health
Grades 6-8, page 145
The environment may contain dangerous levels of substances that are harmful to human beings. Therefore, the good health of individuals requires monitoring the soil, air, and water and taking steps to keep them safe.

NSES Content Standard F Science in Personal and Social Perspectives: Personal Health
Grades 5-8, page 168
Natural environments may contain substances (for example, radon and lead) that are harmful to human beings. Maintaining environmental health involves establishing or monitoring quality standards related to use of soil, water, and air.

Benchmarks 1C The Nature of Science: The Scientific Enterprise
Grades 9-12, page 19
Science disciplines differ from one another in what is studied, techniques used, and outcomes sought, but they share a common purpose and philosophy, and all are part of the same scientific enterprise. Although each discipline provides a conceptual structure for organizing and pursuing knowledge, many problems are studied by scientists using information and skills from many disciplines. Disciplines do not have fixed boundaries, and it happens that new scientific disciplines are being formed where existing ones meet and that some subdisciplines spin off to become new disciplines in their own right.

NSES Content Standard E Science and Technology: Understandings about Science and Technology
Grades 9-12, page 192
Scientists in different disciplines ask different questions, use different methods of investigation, and accept different types of evidence to support their explanations. Many scientific investigations require the contributions of individuals from different disciplines, including engineering. New disciplines of science, such as geophysics and biochemistry often emerge at the interface of two older disciplines.

Benchmarks **4D The Physical Setting: Structure**
of Matter
Grades 9-12, page 80
The nucleus of radioactive isotopes is unstable
and spontaneously decays, emitting particles
and/or wavelike radiation. It cannot be predict-
ed exactly when, if ever, an unstable nucleus will
decay, but a large group of identical nuclei
decay at a predictable rate. This predictability of
decay rate allows radioactivity to be used for
estimating the age of materials that contain
radioactive substances.

NSES Content Standard B Physical Science:
Structure of Atoms
Grades 9-12, page 178
Radioactive isotopes are unstable and undergo
spontaneous nuclear reactions, emitting particles
and/or wavelike radiation. The decay of any one
nucleus cannot be predicted, but a large group of
identical nuclei decay at a predictable rate. This
predictability can be used to estimate the age of
materials that contain radioactive isotopes.

COMPARISON OF *BENCHMARKS* AND
CURRICULUM AND EVALUATION
STANDARDS FOR SCHOOL MATHEMATICS
Unlike the *National Science Education Standards*,
Benchmarks includes detailed mathematics content
integral to science literacy. The *Resources for Science
Literacy: Professional Development* CD-ROM includes
a comparison of *Benchmarks* to the National Council
for Teachers of Mathematics' (NCTM) *Curriculum*

and Evaluation Standards for School Mathematics prepared by a well-known mathematics educator, Gerald
Kulm. In his comparison, there is a brief discussion
of how *Benchmarks* relates to the NCTM *Standards* in
general (see below), then a detailed comparison that
indicates, for each NCTM standard, the benchmarks
that most nearly relate to it. Users of the comparison
can explore both sets of learning goals for mathematics
with either *Benchmarks* or NCTM's *Standards* as their
starting point.

As the comparison of these two documents indicates, *Benchmarks* and the NCTM *Standards* differ
appreciably in domain, purpose, and organization.
Benchmarks includes science, mathematics, and technology and their interconnections, whereas the
NCTM *Standards* deals exclusively with mathematics. And while *Benchmarks* presents learning goals on
the way to adult science literacy, with an emphasis on
what students should *know*, the NCTM *Standards*
are concerned with what students should learn in
school mathematics, with an emphasis on what students should be able to *do*. And, finally, unlike *Benchmarks* the NCTM *Standards* are organized by topic
areas within three grade-level bands, K-4, 5-8, and 9-
12, with somewhat different topic areas at each level.

Given these differences, there is still considerable
agreement between NCTM's *Standards* and *Benchmarks for Science Literacy*. Since NCTM released its
*Curriculum and Evaluation Standards for School Math-
ematics* in 1989 (shortly after the release of *Science for*

All Americans) Project 2061 has endorsed the principles expressed in the *Standards* and subscribed to most of its learning goals. In preparing *Benchmarks for Science Literacy*, the Project took into account the mathematics *Standards*, although usually expressing goals at a more specific level and occasionally having a less ambitious view of what *all* students can learn.

Examples from the CD-ROM

The following material taken from the *Resources for Science Literacy: Professional Development* CD-ROM provides (1) Kulm's broad description of the content areas covered, level of detail, and treatment of content in both NCTM's *Standards* and *Benchmarks* and (2) specific examples of individual benchmarks matched to the relevant NCTM standards.

Overall Content Comparisons

In general, *Benchmarks* tends to provide outlines of the overall literacy expected of students once they finish school, whereas *Standards* states the learning outcomes expected during the study of mathematics in school. *Benchmarks* statements are most often about what students should *know*, whereas statements in *Standards* are about what students should be able to *do*. However, it is clear in *Benchmarks* that the *way* students *become* literate and know is

through doing and investigating. For this reason, some of the *Benchmarks* statements in mathematics may focus on procedures or knowledge about how something works, but imply a deeper understanding and sense of how and why it works.

Benchmarks contains explicit and comprehensive treatment of the nature of mathematics, problem solving, and doing science. *Standards* focuses on specific components of problem solving such as developing strategies and verifying results without explicitly describing the nature of mathematics.

The ideas and processes of reasoning and critical thinking receive a great deal of attention in *Benchmarks*. The ideas of creating arguments and explanations is stressed. Even more important, however, is the notion of being critical of unsound arguments or claims, especially those based on faulty data or faulty reasoning or by persons who are not expert in the field being discussed. In addition, the importance of openness to alternative explanations, careful experimentation, and sound evidence are stressed as features of good science and mathematics.

A great deal of attention is given to statistics in *Benchmarks*, especially their interpretation and cautions about misinterpretation. Although separate treatment of probability computations is not as extensive as in *Standards*, *Benchmarks* has more on the conceptual role of probability in expectation and prediction. Also, some of the statistics benchmarks have implications and interrelationships with probability concepts.

There are some instances of differences in grade-level placement. Algebraic ideas tend to be mentioned by *Benchmarks* in earlier grades than by *Standards*, yet in the later grades, *Benchmarks* treats algebra mainly as computational—stopping short of symbolic manipulation and solution of equations. Graphs, however, are an important topic throughout *Benchmarks*. Computation receives heavy emphasis in *Benchmarks*, being carried into the high school grades. *Benchmarks* often integrates the meaning and understanding of computations, connecting them with applications in familiar contexts. Also, *Benchmarks* provides more concrete examples of how a student's number sense and facility with computations should be demonstrated.

Benchmarks **1B The Nature of Science:**
   Scientific Inquiry
   Grades K-2, page 10
   Tools such as thermometers, magnifiers, rulers, or balances often give more information about things than can be obtained by just observing things without their help.

NCTM Standards

- Standard 10-2, page 51, Grades K-4
  Develop the process of measuring and concepts related to units of measurement
- Standard 13-3, page 116, Grades 5-8
  Select appropriate units and tools to measure to the degree of accuracy required in a particular situation

Benchmarks **2A The Nature of Mathematics:**
   Patterns and Relationships
   Grades 3-5, page 27
   Mathematics is the study of many kinds of patterns, including numbers and shapes and operations on them. Sometimes patterns are studied because they help to explain how the world works or how to solve practical problems, sometimes because they are interesting in themselves.

NCTM Standards

- Standard 3-5, Grades K-4, page 29
  Believe that mathematics makes sense
- Standard 8-3, Grades 5-8, page 98
  Analyze functional relationships to explain how a change in one quantity results in a change in another
- Standard 3-5, Grades 5-8, page 81
  Appreciate the pervasive use and power of reasoning as a part of mathematics

Benchmarks 9A The Mathematical World: Numbers
  Grades 6-8, page 213
  Numbers can be written in different forms, depending on how they are being used. How fractions or decimals based on measured quantities should be written depends on how precise the measurements are and how precise an answer is needed.
NCTM Standards
  - Standard 12-1, Grades K-4, page 57
    Develop concepts of fractions, mixed numbers, and decimals
  - Standard 12-2, Grades K-4, page 57
    Develop number sense for fractions and decimals
  - Standard 4-2, Grades K-4, page 32
    Relate various representations of concepts or procedures to one another
  - Standard 5-4, Grades 5-8, page 87
    Investigate relationships among fractions, decimals, and percents
  - Standard 6-3, Grades 5-8, page 91
    Extend their understanding of whole number operations to fractions, decimals, integers, and rational numbers

Benchmarks 11C Common Themes: Constancy and Change
  Grades 9-12, page 275
  Graphs and equations are useful (and often equivalent) ways for depicting and analyzing patterns of change.
NCTM Standards
  - Standard 4-1, Grades 9-12, page 146
    Recognize equivalent representations of the same concept
  - Standard 5-2, Grades 9-12, page 150
    Use tables and graphs as tools to interpret expressions, equations, and inequalities
  - Standard 6-3, Grades 9-12, page 154
    Translate among tabular, symbolic, and graphical representations of functions
  - Standard 8-1, Grades 9-12, page 161
    Translate between synthetic and coordinate representations
  - Standard 8-2, Grades 9-12, page 161
    Deduce properties of figures using transformations and using coordinates

Comparison of *Benchmarks* and *Curriculum Standards* for *Social Studies*

This brief comparison identifies the extensive overlap of Project 2061's *Benchmarks for Science Literacy*, which includes the social sciences, and the National Council for the Social Studies' (NCSS) *Standards*.

It is important to recognize that Project 2061 does not presume to prescribe learning goals for the social studies curriculum, which already incorporates recommendations from many of the individual disciplines (history, geography, economics, political science, sociology, psychology, civics, etc.). Rather, Project 2061 intends the comparison to be useful to educators in fleshing out the NCSS standards and helpful in promoting further discussion between Project 2061 and the social studies community on similarities, differences, and opportunities for cooperation.

The comparison of *Benchmarks* and the NCSS *Standards* presents a simplified listing of the 80 or so "performance expectations" from the NCSS *Standards*, keyed to relevant chapter sections (not to specific benchmarks) in *Benchmarks*. In the NCSS document, performance expectations are described succinctly (at three grade levels) within the following 10 thematically based curriculum standards:

Culture
Power, Authority, and Governance
Time, Continuity, and Change
Production, Distribution, and Consumption
People, Places, and Environments
Science, Technology, and Society
Individual Development and Identity
Global Connections
Individuals, Groups, and Institutions
Civic Ideals and Practices

For ease of use, the Project 2061 comparison lists these performance expectations in the same order that they appear in the NCSS document and also groups and labels them, based on the NCSS' descriptions of them. Almost half of the NCSS performance expectations have correlates in *Benchmarks*, usually distributed among several *Benchmarks* sections.

Examples from the CD-ROM

Not surprisingly, most of the connections from the NCSS *Standards* are to *Benchmarks* Chapter 7: Human Society, which deals with individual and group behavior, social organizations, and the processes of social change. Yet the comparison also identifies many connections to other *Benchmarks* chapters, as illustrated in the following examples taken from the *Resources for Science Literacy* CD-ROM:

> *NCSS Standards* Individuals, Groups, and Institutions, early grades
> Show how groups and institutions work to meet individual needs and promote the common good, and identify examples of where they fail to do so.
>
> ***Benchmarks* 7E Human Society: Political and Economic Systems**
> **Grades 3-5, page 168**
> - **People tend to live together in groups and therefore have to have ways of deciding who will do what.**

- **Services that everyone gets, such as schools, libraries, parks, mail service, and police and fire protection, are usually provided by government.**
- **There are not enough resources to satisfy all of the desires of all people, and so there has to be some way of deciding who gets what.**
- **Some jobs require more (or more extensive) training than others, some involve more risk, and some pay better.**

NCSS Standards People, Places, and Environments, early grades
Describe and speculate about physical system changes, such as seasons, climate and weather, and the water cycle.

Benchmarks **4B The Physical Setting: The Earth Grades K-2, page 67**

- **Some events in nature have a repeating pattern. The weather changes some from day to day, but things such as temperature and rain (or snow) tend to be high, low, or medium in the same months every year.**
- **Water can be a liquid or a solid and can go back and forth from one form to the other. If water is turned into ice and then the ice is allowed to melt, the amount of water is the same as it was before freezing.**
- **Water left in an open container disappears, but water in a closed container does not disappear.**

NCSS Standards Time, Continuity, and Change, middle grades
Identify and describe selected historical periods and patterns of change within and across cultures, such as the rise of civilizations, the development of transportation systems, the growth and breakdown of colonial systems, and others.

Benchmarks **3C The Nature of Technology: Issues in Technology Grades 6-8, page 56**

- **Throughout history, people have carried out impressive technological feats, some of which would be hard to duplicate today even with modern tools. The purposes served by these achievements have sometimes been practical, sometimes ceremonial.**
- **Technology has strongly influenced the course of history and continues to do so. It is largely responsible for the great revolutions in agriculture, manufacturing, sanitation and medicine, warfare, transportation, information-processing, and communications that have radically changed how people live.**

NCSS Standards Science, Technology and Society, high school
Formulate strategies and develop policies for influencing public discussions associated with technology-society issues, such as the greenhouse effect.

Benchmarks 1C The Nature of Science: The
Scientific Enterprise
   Grades 9-12, page 19
   • **Scientists can bring information, insights,
   and analytical skills to bear on matters of
   public concern. Acting in their areas of expertise, scientists can help people understand the
   likely causes of events and estimate their possible effects. Outside their areas of expertise,
   however, scientists should enjoy no special
   credibility. And where their own personal,
   institutional, or community interests are at
   stake, scientists as a group can be expected to
   be no less biased than other groups are about
   their perceived interests.**

NOTES ON THE COMPARISONS
All of the comparisons included on the *Professional
Development* CD-ROM are meant to help educators
understand the growing consensus among experts on
what students should know in science, mathematics,
and technology and to make better use of *Benchmarks*
and the various *Standards* together or separately. Users
who want to better their general understanding of
learning goals will want to read the "Summary Comparison of Content" document that compares *Benchmarks* to the *National Science Education Standards* and
the "Detailed Content Comparison" section of the
Benchmarks-National Council for Teachers of Mathematics *Standards* comparison (both are accessible on
the CD-ROM). In addition, the *Project 2061 Work-*

shop Guide (also included on the *Professional Development* CD-ROM) offers presentations describing the
comparisons and how to use them effectively. As
educators make choices for their curricula, they are
likely to want to delve more deeply into the detailed
analyses and, at the same time, to consult the original
documents themselves.

FUTURE VERSIONS
The next edition of *Resources for Science Literacy: Professional Development* is expected to include comparisons of Project 2061's learning goals to state- and
district-level documents such as curriculum frameworks and content standards. Project 2061 encourages users to share their materials with us and to provide us with ideas and suggestions for making these
comparisons as useful as possible.
   Send suggestions to:
   Project 2061
   American Association for the Advancement of Science
   1333 H Street, NW
   P.O. Box 34446
   Washington, D.C. 20005
   FAX: 202/842-5196
   Electronic Mail: project2061@aaas.org (please
      identify subject as "comparisons")

For more detailed instructions on using the comparisons, please
refer to Chapter 7: Using the *Resources for Science Literacy*
CD-ROM on page 107 of this volume.

KEITH HARING, *Untitled 1 from Growing Suite, 1988.*

Chapter **6** ABOUT THE *PROJECT 2061 WORKSHOP GUIDE*

With the publication of *Benchmarks for Science Literacy* in 1993, Project 2061 recommended a coherent set of grade-specific learning goals in science, mathematics, and technology for students in kindergarten through 12th grade. These learning goals were derived from the science literacy goals for all high school graduates that Project 2061 had outlined in its 1989 report *Science for All Americans* (*SFAA*).

As educators began to consider the far-reaching implications these goals would have for curriculum, instruction, and assessment, they turned to Project 2061 for help. In response, Project 2061 has developed a variety of workshops for different purposes and audiences, including teachers, teacher educators, administrators, curriculum developers, and policy makers. Over three dozen educators from Project 2061 School-District Center teams and other consultants helped design the presentations and activities and tested them in Project 2061 workshops around the country.

To make this wealth of material widely available and easily accessible, the *Resources for Science Literacy: Professional Development* CD-ROM includes the *Project 2061 Workshop Guide* as one of its six components.

The *Project 2061 Workshop Guide* brings together a wide array of workshop options so that educators can create Project 2061 workshops to suit diverse needs, interests, and time frames. The organization of the *Workshop Guide* on the *Professional Development* CD-ROM is similar to that of a conventional printed volume, with a table of contents and chapters. The following briefly describes the contents and purpose of each chapter:

- **Preparing for a Project 2061 Workshop**
  Here you will find detailed instructions on using the *Workshop Guide* and planning a Project 2061 workshop, including helpful tips for workshop leaders, a workshop planning guide, and information on how to order materials.

- **Designing a Project 2061 Workshop**
  This is the largest section of the *Workshop Guide* on the CD-ROM and contains nearly 60 different

workshop options (i.e., presentations and activities) from which to assemble a customized workshop agenda. Options include suggested scripts, handouts, and transparencies. You can select an option, view it on your computer monitor, and then print all of the materials associated with the option.

- **Handouts**
  This alphabetical listing on the CD-ROM allows you to select, view, and print any of the handouts included in the *Workshop Guide*. Handouts can also be accessed directly from a workshop option.

- **Transparencies**
  This alphabetical listing on the CD-ROM allows you to select, view, and print any of the transparencies included in the *Workshop Guide*. Transparencies can also be accessed directly from a workshop option.

- **Selected Readings for Workshop Leaders**
  This alphabetical listing on the CD-ROM allows you to select, view, and print any of the supplemental readings included in the *Workshop Guide*. Readings can also be accessed directly from a workshop option. Three sample workshop agendas to use as models for designing a Project 2061 workshop are also included in this chapter.

The CD-ROM format of the *Workshop Guide* provides a great deal of flexibility in designing workshops. The *Guide* can also serve as a tutorial for learning more about Project 2061 and how to use its tools for science literacy.

ABOUT PROJECT 2061 WORKSHOPS
All Project 2061 workshops aim to demonstrate to audiences the need for change in science, mathematics, and technology education and to inform them about the help that Project 2061 tools can provide. Workshops generally progress through three major stages—Opening, Project 2061 Tools, and Closing—following the plan shown in **Figure 1**.

Within each stage, there are specific topics that are always addressed in a Project 2061 workshop. The *Workshop Guide* included on the *Professional Development* CD-ROM provides a number of options to address these topics. After deciding on the focus of a workshop, users can select the options that suit the workshop's intended purpose, meet the needs and interests of the participants, and fit the time available. To help prospective workshop leaders get started, the *Workshop Guide* on CD-ROM provides three sample workshop agendas to suggest some of the possibilities for 6-hour, 1.5-day, and 2.5-day workshops. The sample agenda for the 6-hour workshop is also presented in this volume on page 96.

Workshop Options

Each workshop option is a self-contained presentation or activity that can be selected from the *Workshop Guide* on CD-ROM and then combined with other options to plan a workshop agenda. The presentation shown in **Figure 2** is an example of an option that can be used in the Opening stage of a workshop. It is designed to help participants understand why science education reform is needed.

As illustrated in **Figure 2**, each option includes (1) an estimate of the time needed for the option; (2) an example of the option's use; (3) a list of materials and transparencies needed for the presentation; (4) a sample script for the workshop presenter; and (5) cues for showing transparencies and other suggestions for a successful presentation. There is also a brief descriptive overview of each option.

The largest section of the *Workshop Guide* on CD-ROM is Exploring the Use of Project 2061 Tools. This section contains a variety of workshop options that help to describe Project 2061's principles, strategy, and reform tools, and to help participants use *SFAA* and *Benchmarks for Science Literacy* to:

- understand learning goals,
- evaluate curriculum frameworks,
- analyze curriculum materials,
- analyze instruction, and
- improve lesson design.

Typically, a workshop explores only one of these possible uses of Project 2061 tools, depending on the interests and needs of the workshop participants. Additional options for other stages of the workshop— e.g., Introduction, Workshop Goals, Evaluation, and so on—are then selected, and the rest of the workshop is assembled with the Use option as its focus.

Examples from the CD-ROM

To introduce users to the variety of materials available in the *Project 2061 Workshop Guide* on CD-ROM, one of the sample workshop agendas is presented here in its entirety (indicated with a blue vertical line to the left of the text). For each stage of this sample workshop, options have been selected and are described in brief option overviews. Examples of the transparencies, handouts, and supplementary readings associated with each option are also shown in **Figures 3 through 14**. This sample agenda is for a 6-hour workshop designed to help participants find out more about how Project 2061's tools can help them improve teaching.

PLANNING A PROJECT 2061 WORKSHOP AGENDA

Opening

This stage allows the workshop leader to find out more about what participants already know about science education reform and Project 2061 and to establish the specific goals for the workshop.

- Introduction (four options available)
- Need for Change in Science Education (six options available)
- Workshop Goals (four options available)
- What Do Participants Know about Project 2061? (three options available)

Project 2061 Tools

This is the core of a Project 2061 workshop and the option chosen here will help determine which Opening and Closing options are selected.

- Overview of Tools Available from Project 2061 (ten options available)
- Exploring the Use of Project 2061 Tools to (choose one of the following uses):
  - Understand the nature of benchmarks (five options available)
  - Analyze curriculum frameworks (one option available)
  - Analyze curriculum materials (two options available)
  - Analyze instruction (five options available)
  - Design instruction (seven options available)

Closing

In this stage workshop participants reflect on what they have learned and provide the workshop leader with feedback on the effectiveness of the workshop itself.

- Summary (five options available)
- Evaluation (six options available)

Figure 1. Various options from the *Project 2061 Workshop Guide* on CD-ROM can be assembled to build a customized Project 2061 workshop.

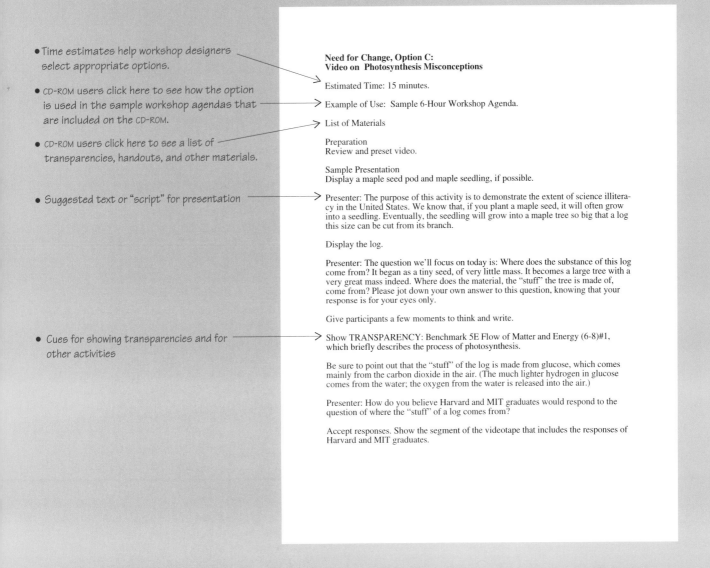

• Time estimates help workshop designers select appropriate options.

• CD-ROM users click here to see how the option is used in the sample workshop agendas that are included on the CD-ROM.

• CD-ROM users click here to see a list of transparencies, handouts, and other materials.

• Suggested text or "script" for presentation

• Cues for showing transparencies and for other activities

Need for Change, Option C:
Video on Photosynthesis Misconceptions

Estimated Time: 15 minutes.

Example of Use: Sample 6-Hour Workshop Agenda.

List of Materials

Preparation
Review and preset video.

Sample Presentation
Display a maple seed pod and maple seedling, if possible.

Presenter: The purpose of this activity is to demonstrate the extent of science illiteracy in the United States. We know that, if you plant a maple seed, it will often grow into a seedling. Eventually, the seedling will grow into a maple tree so big that a log this size can be cut from its branch.

Display the log.

Presenter: The question we'll focus on today is: Where does the substance of this log come from? It began as a tiny seed, of very little mass. It becomes a large tree with a very great mass indeed. Where does the material, the "stuff" the tree is made of, come from? Please jot down your own answer to this question, knowing that your response is for your eyes only.

Give participants a few moments to think and write.

Show TRANSPARENCY: Benchmark 5E Flow of Matter and Energy (6-8)#1, which briefly describes the process of photosynthesis.

Be sure to point out that the "stuff" of the log is made from glucose, which comes mainly from the carbon dioxide in the air. (The much lighter hydrogen in glucose comes from the water; the oxygen from the water is released into the air.)

Presenter: How do you believe Harvard and MIT graduates would respond to the question of where the "stuff" of a log comes from?

Accept responses. Show the segment of the videotape that includes the responses of Harvard and MIT graduates.

Figure 2. An Example of a Workshop Option. Each workshop option is a self-contained presentation or activity which can be combined with other options to form a workshop agenda.

A Sample Agenda for a 6-Hour Project 2061 Workshop

**Exploring the Use of Project 2061 Tools
to Analyze Instruction**

Workshop Summary

This workshop is designed for a group of middle and high school science teachers, many of whom do not think there is a problem with science teaching and learning. Convinced that motivated and able students are learning quite well, these teachers question the need for change.

Using clips from the video *Lessons Pulled from Thin Air*, program 2 in the Annenberg/CPB Project's new Private Universe series, this 6-hour workshop demonstrates that even some graduates of the nation's finest schools do not understand some basic ideas in science. Through a variety of activities and reflection, participants learn about *Science for All Americans (SFAA)* and *Benchmarks for Science Literacy* and how they can be used first to analyze instruction, and then to consider how to improve instruction to help all students achieve the science literacy goals recommended by Project 2061. The estimated time shown for each option is the minimum required.

Opening

Introduction Option C: Teaching Versus Learning Cartoon

Overview: A cartoon (**Figure 3**) is used to focus participants' attention on the gap between what teachers believe they have taught and what students have actually learned. This option can be used to set the stage for a workshop that focuses on exploring the use of Project 2061 Tools to design instruction. Estimated Time: 10 minutes

Need for Change in Science Education Option C: Video on Photosynthesis Misconceptions

Overview: Videotaped interviews (**Figure 4**) are used to show that graduates from the best universities do not under-

stand the idea in the grade 6-8 benchmark, "Plants use the energy from light to make sugars from carbon dioxide and water." This option is particularly appropriate for workshops addressing learning goals found in the Flow of Matter and Energy sections of *SFAA* and *Benchmarks* Chapter 5: The Living Environment. Estimated Time: 15 minutes.

Workshop Goals Option B: Workshop on Teaching

Overview: The presenter shares with the group specific workshop goals (**Figure 5**), which are appropriate for an introductory workshop for teachers. These include making the case that science education is not working well for most students, even in the best schools; understanding that effective teaching should include understanding of science literacy goals; and that study of *SFAA* and *Benchmarks* can improve teaching. Estimated Time: 5 minutes.

What Do Participants Know about Project 2061?

Option A: Participant Questions about Project 2061

Overview: Participants list questions they have about Project 2061, *SFAA*, and *Benchmarks*. The sophistication of their questions indicates to the presenter the extent of their knowledge of Project 2061 and reminds participants that there is a wide range of questions among them. Estimated Time: 10 minutes.

Project 2061 Tools

Tools Available from Project 2061 Option C: Project 2061 Principles, Strategy, and Tools

Overview: Using transparencies and handouts (**Figures 6 and 7**), the presenter provides a brief history of Project 2061 and an overview of its existing products and products to come. The focus is on the systemic, long-term nature of Project 2061 reform. This option works well when only a short amount of time is available, but the audience is unfamiliar with Project 2061. However, if the audience includes more sophisticated users of *Benchmarks*, this option might serve as a quick reminder that Project 2061's broad vision for science education reform encompasses much more than curriculum reform. For workshops of a day or more in length, this option can be combined with Options D or G to provide more detail about *SFAA* and *Benchmarks*. Estimated Time: 20 minutes.

Exploring the Use of Project 2061 Tools, 4: To Analyze Instruction
Option B: Instruction Aimed at Benchmark 5E Flow of Matter and Energy (6-8)#1
Overview: Participants study components of *SFAA*, *Benchmarks*, and *Benchmarks on Disk* (**Figure 8**) to enhance their understanding of the meaning of the grade 6-8 benchmark 5E Flow of Matter and Energy (**Figure 9**). Estimated Time: 3.5 hours.

Participants then read several scenarios (**Figure 10**) describing instruction and evaluate how effectively the instruction seems to address the benchmark (**Figures 11** and **12**). If time permits, participants consider how the instruction could be improved to address the benchmark better.

This option is appropriate for groups that include middle and high school teachers who teach the topic. The strand map (**Figure 13**) shows the importance of precursor benchmarks to this benchmark, the background reading (**Figure 14**) describes strand maps and their purpose, and the scenarios give participants an opportunity to apply their understanding of the benchmark to the analysis and improvement of instruction.

CLOSING
Summary Option C: Participants Reflect on Progress
Overview: Participants identify one workshop goal statement they now believe is true and consider what workshop activities contributed to their conviction. This option is appropriate for any workshop. Estimated Time: 20 minutes.

Evaluation Option A: Using Project 2061 Workshop Evaluation Form
Overview: Participants consider three opinions about *Benchmarks* (**Figure 15**), choose the one they agree with most, and explain their reasoning. This helps the presenter determine whether participants think *Benchmarks* will be useful to their own reform efforts. Estimated Time: 15 minutes.

Figure 3. Transparency used with **Introduction** Option C

Figure 4. This clip from an Annenberg/CPB Project video used in the **Need for Change** Option C portrays the extent of science illiteracy in the U.S.

Points to Be Made

■ Science education is not working for most students, even in the best schools.

■ A thorough understanding of science literacy and learning goals is essential for effective teaching.

■ Use of *SFAA* and *Benchmarks* can promote more effective teaching.

Figure 5. Transparency used with **Workshop Goals** Option B lists the specific objectives for the workshop.

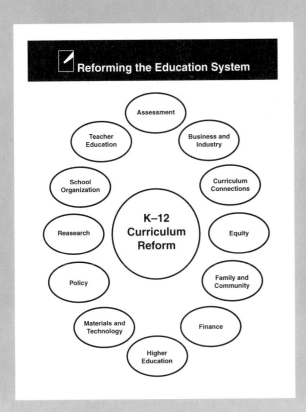

Figure 6. Transparency used with **Tools Available from Project 2061** Option C depicts the systemic nature of Project 2061 reform.

Figure 7. Handout used with **Tools Available from Project 2061** Option C describes Project 2061.

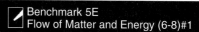

Understanding the Nature of a Benchmark: A Study Procedure

An important part of every Project 2061 workshop is helping participants study Project 2061 tools in a way that increases their understanding of benchmarks. Benchmarks tend to be interpreted within the context of what is currently being taught or what teachers themselves learned in school, and participants often overestimate or underestimate the extent of learning that is intended in a benchmark. As a result of seeing how often educators jump to conclusions about what benchmarks mean, Project 2061 staff have included in their workshops a procedure in which participants are asked to read and discuss with one another what a particular benchmark is expecting students to know or be able to do. Then they are asked to examine other parts of *Science for All Americans* (*SFAA*) and *Benchmarks for Science Literacy* (*Benchmarks*) that can shed light on the benchmark's intent. Participants read and discuss five items:

The *SFAA* section from which the benchmark originated.
SFAA recommends goals for adult science literacy; *Benchmarks* recommends specific learning goals for grades 2, 5, 8, and 12 that can contribute to adult literacy. There is a corresponding *Benchmarks* section for each *SFAA* section. Reading the *SFAA* section helps participants understand what literacy in that topic is defined to be, and thus where the benchmark is aiming.

All other benchmarks from the K-12 list of benchmarks in the same *Benchmarks* section.
Reading the other benchmarks helps participants understand the level of sophistication intended by the benchmark.

Introductory essays in the *Benchmarks* section for the benchmark being studied.
The section introduction and grade-level essays help participants understand difficulties students may have with the benchmark topic. They also offer some suggestions for helping students achieve the benchmark.

Summaries of research on the topic from *Benchmarks*, Chapter 15.
The research selection suggests likely limitations in student understanding of the benchmark (and therefore its grade placement) and points participants to the original research articles.

A relevant strand map from *Benchmarks on Disk*.
A strand map helps participants see how other benchmarks relate to the benchmark being studied and their importance for understanding that benchmark.

Benchmark 5E
Flow of Matter and Energy (6-8)#1

Food provides molecules that serve as fuel and building material for all organisms. Plants use the energy from light to make sugars from carbon dioxide and water. This food can be used immediately or stored for later use. Organisms that eat plants break down the plant structures to produce the materials and energy they need to survive. Then they are consumed by other organisms.

Figure 8. Handout used with **Exploring the Use of Project 2061 Tools 4** Option B describes a systematic procedure for studying specific benchmarks to clarify their full meaning.

Figure 9. Transparency used with **Exploring the Use of Project 2061 Tools 4** Option B presents the grade 6-8 benchmark on the flow of matter and energy. Workshop participants use this benchmark as the basis for analyzing a set of instructional scenarios.

Instructional Scenarios for Benchmark 5E (6-8) #1

Instructional Scenario A

"Today we will do an experiment to find out whether light is necessary for photosynthesis," Mrs. Goodman told her 7th-grade class. "We've just learned that, in photosynthesis, plants make sugar, which then turns into starch. Let's find out whether light is necessary for that process."

She gave each group of four students a geranium plant, as well as written directions for the experiment. Carefully following the directions, each group covered a leaf with aluminum foil. The students then watered the plants.

"Where shall we put them?" asked Roberta.

"Where would you suggest?" responded Mrs. Goodman.

"I think they should go on the window sill, where they'll get lots of light," suggested Bryan.

"Good idea," said Mrs. Goodman. "What would be evidence that light **is** required for photosynthesis?" she asked.

After thinking a moment, Bryan said, "We'd have to find sugar or starch in the leaves that were uncovered so they got light."

"And what would be evidence that light **is not** required for photosynthesis?" asked Mrs. Goodman.

"If we could put a plant in the dark and it did make sugar," said Brent.

"If the leaf we've covered did make sugar or starch," suggested Roberta.

The next day the students practiced testing various items for the presence of starch. They found that an iodine solution turned dark blue, almost black, in the presence of starch, but the iodine solution remained light brown if no starch was present.

The fourth day after they had covered the leaves, the students continued the experiment. Mrs. Goodman donned her safety goggles and gave each student a pair. "Put on your goggles before you work with the hot liquids," she said. The students put on their goggles and set to work. They removed an uncovered leaf from each plant, placed it in boiling water for a few minutes, and then dipped it in a hot alcohol solution until the green color was gone. They then tested it with iodine solution. All the leaves turned blue-black.

Next the students tested the covered leaf from each plant, using the same procedures. They were very careful, for example, to place this leaf in boiling water for the same length of time as the uncovered leaf had been in boiling water. When they tested the leaves that had been covered for the presence of starch, almost all students noted that the iodine solution remained light brown.

In one case, a blue-black splotch appeared on a leaf.

"That proves you can have photosynthesis even without light," said Alex.

"Wait," said Amy. "I was in that group, and I was the one who unwrapped that leaf. I noticed a little tear in the foil. So maybe some light got in through the tear."

"You'd better repeat the experiment," suggested Alex.

Pedagogical Match Questions
Analysis of Instruction

Are opportunities provided for teachers to find out what students already think about the ideas in a benchmark at the beginning and throughout the instruction? Is the information used?

Are students engaged in activities (including reading and listening to peers and the teacher) and provided with opportunities to reflect on their activities?

Are students given experiences with concepts before terms are introduced?

Are students engaged in problems and questions before they are introduced to ideas or solutions?

Does the scenario describe instruction that helps students make connections between benchmarks and their prerequisites?

Figure 10. Instructional scenarios handout used with **Exploring the Use of Project 2061 Tools 4** Option B portrays several different ways of teaching and learning about specific ideas in *Benchmarks*.

Figure 11. Transparency used with **Exploring the Use of Project 2061 Tools 4** Option B presents questions about pedagogy that help workshop participants analyze the effectiveness of instruction.

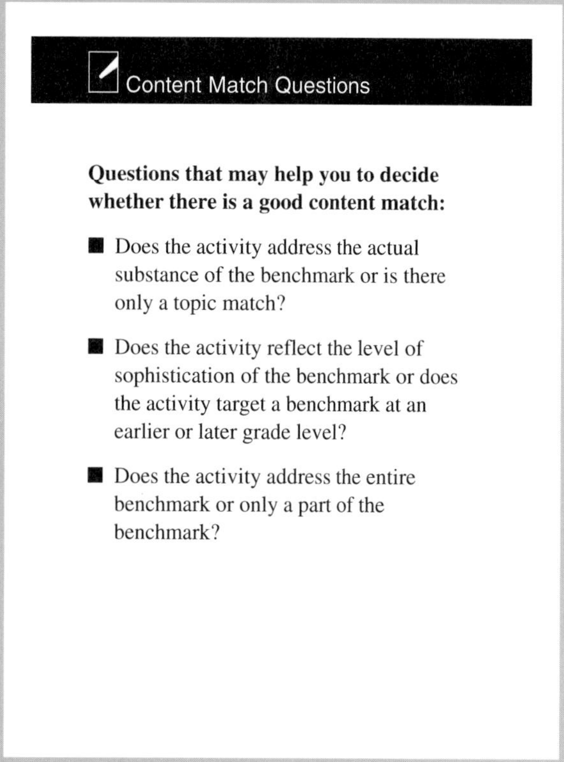

Content Match Questions

Questions that may help you to decide whether there is a good content match:

- Does the activity address the actual substance of the benchmark or is there only a topic match?

- Does the activity reflect the level of sophistication of the benchmark or does the activity target a benchmark at an earlier or later grade level?

- Does the activity address the entire benchmark or only a part of the benchmark?

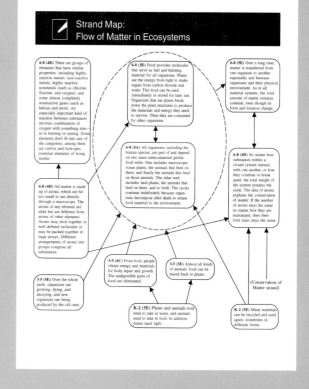

Strand Map:
Flow of Matter in Ecosystems

Figure 12. Transparency used with **Exploring the Use of Project 2061 Tools 4** Option B lists questions about content that help workshop participants analyze the effectiveness of instruction.

Figure 13. Transparency of a strand map shows how students' understanding of the flow of matter in ecosystems progresses over time. This transparency is used with **Exploring the Use of Project 2061 Tools 4** Option B.

About Strands

Strands, or strand maps, are networks of benchmarks through which students might progress on their way to the adult literacy goals defined in *Science for All Americans*. The strands show the development of concepts from rudimentary benchmarks at the elementary level through middle-school learning to the sophisticated level of understanding expected of high school graduates. Strand maps do not appear in the book version of *Benchmarks for Science Literacy*, but 30 sample strand maps are included on *Benchmarks on Disk*. Strands can be used in the analysis and planning of a K-12 curriculum and can inspire users to develop additional strands for other goals in *Science for All Americans*.

The strand maps on *Benchmarks on Disk* do not show the text of the included benchmarks, however. Using graphics software, workshop leaders can create flowchart maps that include benchmarks text (see, for example, the **Water Cycle** strand). These graphical strand maps better illustrate the interdependence of benchmarks across the grade-span, sections, and topics. They help participants understand how individual benchmarks contribute to progressive achievement of science literacy by showing how related benchmarks build on and reinforce one another. Seven strand maps are included in this *Guide*.

As you examine strands you will see that sometimes a later benchmark explains an earlier one or brings an earlier concept to a more sophisticated level. Sometimes a benchmark provides an example of a generalization in another benchmark, thus both extending and reinforcing it. Sometimes two or more ideas converge at a higher grade level to form a more complex idea, or they reinforce one another.

Note that the language in the collection of benchmarks is not thoroughly fine-tuned—that is, any one benchmark was not conscientiously tuned to relate optimally to every other related benchmark, especially if the related benchmarks are in different chapters. The process of drawing strand maps reveals these imperfect fits, and subsequent versions will be improved. Project 2061 encourages you to share any new strand maps you develop or any suggestions you have for improving the sample strand maps included on the disk.

Features of Strand Maps on *Benchmarks on Disk*

The title identifies the *Science for All Americans* topic addressed by the strand.

A code indicates the chapter, section, and benchmark sequence in *Benchmarks for Science Literacy*.

Figure 14. Participants find out more about progression-of-understanding maps in the background reading selection "About Strands," which is used with **Exploring the Use of Project 2061 Tools 4** Option B.

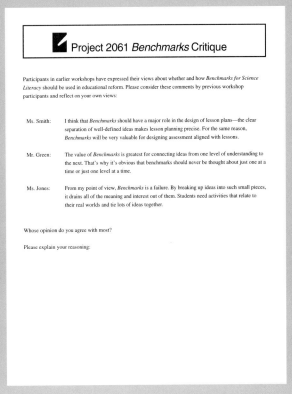

Figure 15. Handout used with **Evaluation** Option A requests feedback from workshop participants about the workshop itself and about Project 2061.

NOTES ON THE *PROJECT 2061 WORKSHOP GUIDE* ON CD-ROM

To make it easy to browse through the abundance of material gathered together in the *Project 2061 Workshop Guide* on CD-ROM, users can preview and select appropriate options directly from lists of options for each stage of the workshop.

Educators who want to conduct their own workshop will find it helpful to begin their study of the *Workshop Guide* on CD-ROM with the section on Preparing for a Project 2061 Workshop before moving on to the workshop options presented in the Designing a Project 2061 Workshop section. This information will also be useful to educators who want to enhance their own understanding of Project 2061 and its reform tools.

Although the *Project 2061 Workshop Guide* on CD-ROM contains a vast collection of materials, its many electronic links allow users to move easily and quickly among options, transparencies, handouts, readings, and sample agendas and from one part of the *Guide* to another.

FUTURE VERSIONS

The Project 2061 *Workshop Guide* will be revised and updated based on feedback from workshop presenters and participants. Project 2061 welcomes suggestions for improving its workshop activities and materials.

Send suggestions to:
Project 2061
American Association for the Advancement of Science
1333 H Street, NW
P.O. Box 34446
Washington, D.C. 20005
FAX: 202/842-5196
Electronic Mail: project2061@aaas.org (please
  identify subject as "workshop")

ALSO SEE ☞ For more detailed instructions on using *Project 2061 Workshop Guide* on CD-ROM, please refer to Chapter 7: Using the *Resources for Science Literacy* CD-ROM on page 107 of this volume.

INSTALLATION

To install the *Resources for Science Literacy* application, you must have the following hardware and software capacity:

- IBM or compatible computer system running Microsoft Windows 3.1® or Microsoft Windows 95®
  OR
- Macintosh® computer running System 7.5 or higher
- 2X or faster CD-ROM reader
- Color monitor is preferred (SVGA at 256 colors is best).

Microsoft Windows 95® and Windows 3.1®. The *Resources for Science Literacy: Professional Development* application runs from the CD-ROM; if you would like to add it to your start menu (in Windows 95®) or your program manager (in Windows 3.1®), double-click on the file labeled **winsetup.exe** on the application CD. This will also install the Adobe Acrobat® Reader which is necessary for printing. Please note, however, that the CD-ROM will be required each time you use the application.

Macintosh®. To install the *Resources for Science Literacy: Professional Development* application, you must first install the Adobe Acrobat® Reader (required for printing) and then copy a small launcher to your hard drive:

- Installing Adobe Acrobat® Reader
  Double-click on the *Resources for Science Literacy: Professional Development* CD icon, then double-click to open the folder labeled **acromac.** Open the folder inside labeled **disk 1** by double-clicking and then double-click on the file **acroread.mac** to install the reader. On-screen instructions will direct the installation from this point on.
- Copying the Launcher to your Hard Drive
  Although the *Resources for Science Literacy: Professional Development* application runs from the CD-ROM, its operation will be enhanced by copying the file marked **rsl-pd launcher** to your hard drive. To do so, drag the file from the CD icon to your desktop or hard drive icon. A message will ask you to confirm your wish to copy this file from the read-only medium. Once you've copied the file, double-click on it to begin using the application. The CD-ROM is required each time you use the application.
- Troubleshooting Applescript™ Extension
  If you do not have the Applescript™ extension active, you will receive an error message. Applescript™ ships with System 7.5 and later, so if it's not in your extensions folder (located inside the system folder on your hard drive), use the **find** feature on the **file** menu to locate it and drag it into the extensions folder. Restart your computer to reactivate the extension.

GETTING STARTED

Starting the Application. To open the application, double-click on the application icon. For Microsoft Windows 95® systems the icon appears under the *Resources for Science Literacy* label on the start menu. For Microsoft Windows 3.1® systems the icon appears in the *Resources for Science Literacy* window. For Macintosh® systems, the icon is labeled **rsl-pd launcher**.

Minimizing. The application window size is fixed (it cannot shrink or grow); however, you may find it convenient to minimize the window at times. Minimizing does not close the application, but the window is temporarily represented on your desktop or toolbar as an icon. To minimize:

- **Microsoft Windows 3.1®.** In the upper right corner of the window, click once on the small box with the downward-pointing triangle. The application, which is still running, is now inactive and represented by an icon on your desktop. Double-click on the icon to reactivate the program.
- **Microsoft Windows 95®.** In the upper right corner of the window, click once on the small box with the black rectangle. The application, which is still running, is now inactive and represented by a labeled rectangle on your toolbar. Click on the rectangle to reactivate the program.
- **Macintosh®.** In the upper right corner of your screen, click and hold on the small icon. Select **Hide *Resources for Science Literacy***. To reactivate the program, click and hold on the same icon and select **Show *Resources for Science Literacy***.

Screen Resolution. *Resources for Science Literacy* is designed to run at the standard 640 x 480 screen size. If your screen resolution is higher, the application window will not fill the entire screen.

Exiting the Application. At any point, you may exit *Resources for Science Literacy* (close the application) by selecting **quit** from the **file** menu, which is the first entry on the toolbar along the top of the application window.

Using the Menu Bar. Several commands are available at all times throughout the application via the toolbar located at the top of the window.

- **File.** The single choice presented in the **file** menu is **quit**. Select this when you wish to close the application, ending your *Resources for Science Literacy* session.
- **Navigate.** This menu lists the six volumes included in *Resources for Science Literacy*. Use it to go immediately to the initial menu for the selected volume.
- **Help.** The **about** section of **help** displays information about the application developers. For help in using the application or for more information about the various components (or volumes) presented on the *Resources for Science Literacy* CD-ROM or about Project 2061, select the **help** icon located at the lower left corner of the application's initial screen.
- **Print.** The print menu only appears when you are viewing an item which can be printed (everything but the menus can be printed). **Print** is the only choice presented in this menu. Once you select it, a print dialog box will appear from which you will need to confirm the print command by pressing the **ok** button.

EXPLORING THE RESOURCES

The initial screen for *Resources for Science Literacy* is a stack of books, each representing one of the six components (or volumes) presented on the CD-ROM. Simply click on a volume to open it. An icon available on most screens allows you to return to this starting point in the application. The **help** button on this screen provides a general overview of the components as well as guidance on navigating the CD-ROM and making use of specific features. Throughout the CD-ROM, click on underlined and colored text to access additional information.

Science for All Americans. The full text of Project 2061's 1989 report *Science for All Americans* is available on the CD-ROM. An initial menu screen lists the 15 chapters of *Science for All Americans*. Select a chapter, and a menu of the sections within that chapter will appear. Select a section to begin reading the text. To advance to the next section, click on the small arrow pointing to the right on the toolbar below the text. The left-pointing arrow buttons take you back to the previous topic, the chapter-level menu, or the initial table of contents for the entire report. As you move your mouse over each button, the text box in the lower right corner of the window describes the button's function.

Science Trade Books. This component provides a descriptive review, table of contents, and bibliographic information for 120 highly-recommended books written for the general reader on all areas of science, technology, and mathematics. The first menu screen allows you to select books by title, by author, or by their relevance to specific chapters and sections of *Science for All Americans.*

Books are arranged alphabetically in separate lists by title and by author. To select books that relate to specific ideas found in *Science for All Americans*, click on the chapter or section in which you are interested.

Each book screen reproduces the book's cover and provides bibliographic data and a review or description of the book. For information on how to contact the publisher, click on the publisher's name (in red). This information will also appear as a footnote when you print the book screen. To access the book's table of contents, use the toggle button below the text window; use this same button to return to the book screen. Both the book screen and the contents screen can be printed. The arrow button takes you back to the listing by book title, author, or topic where your search began.

Cognitive Research. This component guides users to information on research materials that relate to many of the ideas found in *Science for All Americans*. The first menu screen allows you to select how you would like to review the available research: alphabetically by title or author or topically by *Science for All Americans* chapter. If you choose to search this component's database by topic, the next menu lists the types of research available: articles, books, reports, videos, and the research citations found in Project 2061's *Benchmarks for Science Literacy*, Chapter 15: The Research Base. A list of all the research of the type selected will appear. From that list you may then choose to view specific research citations. Each citation includes an abstract and information about obtaining the full text of the original research. The complete citation can be printed as well as viewed.

College Courses. Detailed information about 15 undergraduate courses designed to promote science literacy are presented in this component. Each course has been analyzed and organized into as many as eight sections. An initial menu lists the courses in alphabetical order by author. After selecting a course to view, you can then choose from the syllabus outline which section you would like to explore. This outline will vary from course to course, depending on the information available. All courses include a detailed syllabus, information about the author, and a list of topics in *Science for All Americans* related to the course material. Use the arrow button to take you back to the course outline and back to the list of all courses in the database.

Comparisons of *Benchmarks* to National Standards. Content standards for mathematics, science, and social studies are compared to the learning goals recommended in Project 2061's *Benchmarks for Science Literacy*. The first menu allows you to select which set of comparisons you wish to view. Next you can choose to view a summary or introduction to each comparison or go directly to the comparison itself. The mathematics and science comparisons can be viewed in either benchmarks or standards order. Use the arrows to move to the benchmark or standard that precedes or follows the one you are viewing. Use the double arrows to return to previous menus.

Project 2061 Workshop Guide.. The *Guide* provides workshop organizers with the background information and presentation materials they need to design and conduct a variety of Project 2061 workshops to suit different audiences and purposes. The basic unit of the *Guide* is a workshop option which typically includes a suggested script for the workshop, a list of transparencies and handouts needed, and background information for the workshop leader. As you use the *Guide*, click on underlined text that appears on the screen in a different color to access other, related workshop elements. The workshop guide is organized into five chapters:

- Chapter 1: Preparing for a Project 2061 Workshop is a good starting point for using the *Workshop Guide*. It includes "How to Use This *Guide*," which can be printed (as can all the background text, presentations, handouts, transparency masters, and readings) for reference while using the *Guide*.
- Chapter 2: Designing a Project 2061 Workshop contains more than 60 options that can be combined to create many different kinds of workshops. Choose "Overview" to read a brief description of an option or click on the option title to open it directly. You can open handouts, transparency masters, and associated readings via the option in which they are used, or you can select them from the alphabetical listings in Chapters 3, 4, and 5. Use the double arrow to return to the option and to the *Workshop Guide* table of contents.

- Chapter 3: Handouts is an alphabetical listing of all handouts available in the *Workshop Guide*. A representation of each handout appears on the screen for review before printing. Use the arrow to advance through each handout one page at a time; the entire handout will be printed when you select the "Print" button. The arrow returns you to your place in the alphabetical listing.
- Chapter 4: Transparencies is an alphabetical listing of all transparencies available in the *Workshop Guide*. As with handouts, you can view a transparency before printing it, and the arrow returns you to your place in the alphabetical listing.
- Chapter 5: Selected Readings for Workshop Leaders is an alphabetical listing of all background and supplementary materials—sample agendas, essays, and notes—that are included in the *Workshop Guide*. Click on the material to view it before printing. Use the arrow to return to your place in the alphabetical listing.

FOR MORE INFORMATION

Project 2061 welcomes your feedback on the *Resources for Science Literacy: Professional Development* CD-ROM and encourages users to suggest ways to improve it. Please send your comments and questions to:

Project 2061
American Association for the Advancement of Science
1333 H Street, NW
P.O. Box 34446
Washington, D.C. 20005
TELEPHONE: 202/326-7002
FAX: 202/842-5196
Electronic Mail: project2061@aaas.org
(please identify subject as "rsl-cd")

ACKNOWLEDGMENTS

The individuals listed here have contributed their time and expertise to the creation of *Resources for Science Literacy: Professional Development*. Whether they recommended titles for the trade books database, field-tested a workshop presentation, developed a course, or made any of a number of other important contributions, their efforts have been essential to our work.

Joan Abdallah *Howard County Public Schools, Maryland*

Ethan Allen *Teachers Academy for Mathematics and Science, Chicago, Illinois*

Nancy Armour *Beaver Dam Elementary School, Georgia*

Myra Baltra *Decatur Elementary School, Pennsylvania*

Anne Batey *Northwest Regional Educational Laboratory, Portland, Oregon*

Charles Beehler *Science Teaching Center, Widener University, Pennsylvania*

Jerry A. Bell *Directorate for Education and Human Resources, American Association for the Advancement of Science*

Ann Benbow *American Chemical Society, Washington, D.C.*

Bob Borjes *Peterson Middle School, Texas*

Alfred B. Bortz *Department of Elementary, Secondary, and Reading Education, Duquesne University*

Jane Bowyer *Department of Education, Mills College*

Donna Brearley *Discovery Place Incorporated, North Carolina*

Merle Bruno *School of Natural Science, Hampshire College*

Stephen Brush *Institute for Physical Science and Technology, University of Maryland*

Chris Bull *Division of Engineering, Brown University*

Rodger Bybee *National Research Council, Washington, D.C.*

Barbara Carroll *La Center Intermediate School, Washington*

Christina Castillo-Comer *Longfellow Middle School, Texas*

Dennis Cheek *Department of Education, Rhode Island*

C. David Christensen *Department of Curriculum and Instruction, University of Northern Iowa*

Liz Clark *Isely &/or Clark Design, Greenbelt, Maryland*

Cheryl Cliett *SERVE Consortium for Mathematics and Science Education, Florida*

Marilyn Cook *H. G. Olsen Elementary School, Texas*

Suzanne Corter *Upper Darby School District, Pennsylvania*

Dennis Coutu *West Bay Collaborative, Rhode Island*

Chris Demers *Dr. H. O. Smith Elementary School, New Hampshire*

Denise Denton *Department of Electrical and Computer Engineering, University of Wisconsin, Madison*

Marcia Denton *Peterson Middle School, Texas*

Paul Divucci *Duquesne University*

Joan Duea *Professor of Education, University of Northern Iowa*

Sarah Duff *Institute for Mid-Grades Reform, Baltimore City Schools*

Bill Duff *School for the Arts, Baltimore City Schools*

Bill Duffy *Columbia Township Schools, Indiana*

Jim Dunlap *Los Angeles Unified School District*

Dianne Erickson *Department of Science and Math Education, Oregon State University*

Debbie Faigenbaum *San Francisco Community Alternative School*

Eileen Ferrance *Department of Education, University of Rhode Island*

Ruth Fortney *Indian Mound Middle School, Wisconsin*

Joseph Foster, Jr. *Leeds Middle School, Pennsylvania*

Linda Froschauer *Weston Middle School, Connecticut*

Alejandro Gallard *Science Education Unit, Florida State University*

April Gardner *Education Division, University of Northern Colorado*

Tracy Gath *Editor,* Science Books & Films

Robert Gauger *Oak Park and River Forest High School, Illinois*

Ethel Gilliam *Ft. Corry Elementary School, Georgia*

Timothy Goldsmith *Department of Biology, Yale University*

Wilma Guthrie *Hewitt-Trussville Junior High School, Jefferson County School District, Alabama*

Mark Hafner *Department of Zoology and Physiology, Louisiana State University*

Penny Hamrich *Curriculum, Instruction and Teaching in Education, Temple University*

Peg Hanley *Eisenhower National Clearinghouse, Columbus, Ohio*

Karen Hardy *Blackwell Elementary School, Georgia*

Barrett Hazeltine *Division of Engineering, Brown University*

Henry Heikkinen *Department of Chemistry and Biochemistry, University of Northern Colorado*

Marlene Hilkowitz *Office of Curriculum Support, School District of Philadelphia*

Ivy Hill *Baltimore City Public Schools*

Chris Holle *Los Angeles Systemic Initiative*

Nicole Holthuis *School of Education, Stanford University*

Gerald Holton *Department of Physics, Harvard University*

Chuck Hunt *Clarke Middle School, Georgia*

John Isely *Isely &/or Clark Design, Greenbelt, Maryland*

Joan Jordan *Elbert County Middle School, Georgia*

Janet Kegg *Librarian, American Association for the Advancement of Science*

Saundra Kent *McKelvie Middle School, New Hampshire*

Tricia Kerr *Kentucky State Department of Education*

Steven Kornguth *University of Wisconsin, Madison*

Nonnie Korten *Monlux MST Center, North Hollywood, California*

Melvin Krantzberg *School of History, Technology, and Society, Georgia Institute of Technology*

Paul Kuerbis *Education Department, Colorado College*

Gerald Kulm *American Association for the Advancement of Science*

John Layman *Department of Physics, University of Maryland*

Norman Lederman *Science and Mathematics Education, Oregon State University*

Elliott Lewis *Department of Health Sciences, Franklin Learning Center High School, Pennsylvania*

Tom Liao *Department of Technology and Society, State University of New York/Stonybrook*

Monica Lochner *Conrad Elzehjem School, Wisconsin*

Mary Lowe *Department of Physics, Loyola College*

Nancy Lowry *School of Natural Science, Hampshire College*

Sharon Lynch *Department of Teacher Education and Special Education, George Washington University*

Colin Mably *Evergreen Communications, London*

Sue Matthews *Elbert County Middle School, Georgia*

Lillian McDermott *Department of Physics, University of Washington, Seattle*

Bruce McGirr *Marston Middle School, California*

Michael McKinney *The Gryphon Group*

Roger Mecouch *Upper Darby School District, Pennsylvania*

Marilyn Melstein *Office of Educational Technology, School District of Philadelphia*

Carolyn Minor *Office of Curriculum Support, School District of Philadelphia*

Jim Minstrell *Assessment, Curriculum, and Teaching Systems for Education, Washington*

Joseph E. Moore, Jr. *Richmond County School System, Georgia*

Kathleen O'Sullivan *Department of Secondary Education, San Francisco State University*

Robert Pollack *Department of Biology, Columbia University*

Harold Pratt *Biological Science Curriculum Study, Colorado Springs, Colorado*

Philip Regal *Department of Zoology, University of Minnesota*

Rita Rice *Office of Schools, School District of Philadelphia*

Dorene Rojas Medlin *Dougherty County Board of Education, Georgia*

Pat Rossman *Conrad Elvehjem School, Wisconsin*

Barbara Salyer *Southwest Educational Development Laboratory, Texas*

Kate Scantlebury *Department of Chemistry and Biochemistry, University of Delaware*

Mark Schug *Department of Curriculum and Instruction, University of Wisconsin, Madison*

Ethel Schultz *The Noyce Foundation*

Judy Schwartz *Lutherville Elementary School, Maryland*

Lana Scott *Indian Mound Middle School, Wisconsin*

Marcia Scott *College Park, Maryland*

Ezra Shahn *Department of Biological Sciences, Hunter College of the City University of New York*

Robin Sharp *San Francisco Community Alternative School*

Rick Silverman *Education Department, University of Northern Colorado*

Lynette Smith *Wanamaker Middle School, Philadelphia*

Mike Smith *Education Department, University of Delaware*

Cary Sneider *Lawrence Hall of Science, University of California, Berkeley*

Susan Snyder *Teacher Enhancement, National Science Foundation*

Maria Sosa *Editor-in-Chief*, Science Books & Films

Sherry A. Southerland *Graduate School of Education, University of Utah*

Barbara Spector *University of South Florida*

Roger L. Spratt *Mesa Public Schools, Arizona*

Victor Stanionis *School of Arts and Science, Iona College*

Carol Stearns *Merck Institute for Science Education*

Anita Stockton *Baltimore County Public Schools*

Irene C. Swanson *Los Angeles Systemic Initiative, Los Angeles Unified School District*

Cheryl Granade Sullivan *Educational Consultant, Georgia*

Gloria J. Takahashi *La Habra High School, California*

Carol Takamoto *Los Angeles Unified School District*

Ernie Thieding *Indian Mound Middle School, Wisconsin*

Domenic Thompson *Baltimore City Public Schools*

Ken Tobin *Science Education Unit, Florida State University*

Clara Tolbert *Office of Leadership and Teaching, School District of Philadelphia*

Jerold Touger *Department of Natural Science and Mathematics, Curry College*

Jan Tuomi *National Research Council, Washington, D.C.*

Toni Ungaretti *Education Division, Johns Hopkins University*

Lisa Usher *Monlux MST Center, North Hollywood, California*

JoAnne Vasquez-Wolf *Mesa Public Schools, Arizona*

Susan Wachowiak *Old Town Program School, California*

Patricia Wang-Iverson *Research for Better Schools, Philadelphia, Pennsylvania*

Sam Ward *Department of Molecular and Cell Biology, University of Arizona*

Kenneth Welty *Communication, Education, and Training, University of Wisconsin, Stout*

Wanda White *Georgia Initiative in Mathematics and Science*

Paul H. Williams *University of Wisconsin, Madison*

David Wong *National Center for Research on Teacher Learning, Michigan State University*

Susan Yoder *Editorial Consultant, Washington, D.C.*

Page x, Photograph courtesy of the National Gallery of Art, Washington, D.C., gift of Jo Ann and Julian Ganz, Jr., 1990.60.1.

Page 12, Photograph courtesy of Allan Stone Gallery, New York, NY.

Page 17, Cover reprinted, by permission, from *Great Essays in Science*, edited by Martin Gardner, Prometheus Books, 1994.

Page 19, Cover reprinted, by permission, from *I Want to Be a Mathematician: An Automathography* by Paul R. Halmos, Springer-Verlag, 1985.

Page 21, Cover reprinted, by permission, from *To Engineer is Human* by Henry Petroski, Random House, 1982.

Page 23, Cover reprinted, by permission, from *A Physicist on Madison Avenue* by Tony Rothman, Princeton University Press, 1991.

Page 25, Cover reprinted, by permission, from *The Beak of the Finch: A Story of Evolution in Our Time* by Jonathan Weiner, Alfred A. Knopf Inc., 1994.

Page 27, Cover reprinted, by permission, from *The Ascent of Man* by Jacob Bronowski, Little, Brown, 1976.

Page 29, Cover reprinted, by permission, from *Metaman: The Merging of Humans and Machines into a Global Superorganism*, Simon & Schuster, 1993.

Page 31, Cover reprinted, by permission, from *So Shall You Reap: Farming and Crops in Human Affairs* by Otto T. Solbrig and Dorothy J. Solbrig, Island Press, 1994.

Page 33, Cover reprinted, by permission, from *On the Shoulders of Giants: New Approaches to Numeracy*. Copyright 1990 by the National Academy of Sciences, National Academy Press.

Page 35, Cover reprinted, by permission, from *The History of Science from 1895 to 1945* by Ray Spangenburg and Diane K. Moser, Facts on File, 1994.

Page 37, Cover reprinted, by permission, from *Diatoms to Dinosaurs: The Size and Scale of Living Things* by Chris McGowan, Island Press, 1994.

Page 39, Cover reprinted, by permission, from *The Science Gap: Dispelling the Myths and Understanding the Reality of Science* by Milton A. Rothman, Prometheus Books, 1991.

Page 42, Photography courtesy of Albert Paley Studios.

Page 60, Photograph courtesy of the Henry Art Gallery, University of Washington, gift of the artist, 70.5.

Page 74, Photograph by Steve Lopez, courtesy of Louis K. Meisel Gallery, New York, NY, #6547.

Page 90, Copyright The Estate of Keith Haring.

Page 97, TIGER, reprinted with special permission of King Features Syndicate.

Page 98, Figure 4, Courtesy of the Private Universe Project. Copyright © President and Fellows of Harvard College.

Page 106, Illustration by Noah Dan.

American Association for the Advancement of Science. *Benchmarks for Science Literacy*. New York: Oxford University Press, 1993.

American Association for the Advancement of Science. *The Liberal Art of Science: Agenda for Action*. Washington, D.C.: Author, 1990.

American Association for the Advancement of Science. *Science for All Americans*. New York: Oxford University Press, 1990.

Gallant, Roy A. *Young Person's Guide to Science: Ideas That Change the World*. New York, Macmillan, 1993.

Hazen, Robert M. and James Trefil. *Science Matters: Achieving Scientific Literacy*. New York: Doubleday, 1991.

National Council for the Social Studies. *Curriculum Standards for Social Studies*. Washington, D.C.: Author, 1994.

National Council of Teachers of Mathematics. *Curriculum and Evaluation Standards for School Mathematics*. Reston, VA: Author, 1989.

National Research Council. *National Science Education Standards*. Washington, D.C.: National Academy Press, 1996.

Rensberger, Boyce. *How the World Works: A Guide to Science's Greatest Discoveries*. New York: Morrow, 1986.

Notes

NOTES

NOTES

NOTES

NOTES

NOTES

Oxford University Press
End-User License

All material recorded on the enclosed media (the software) and in the accompanying manual is copyright © by the American Association for the Advancement of Science (AAAS) and published by Oxford University Press, Inc. (OUP). The software is proprietary to OUP or its suppliers (if any) as set forth in the software and the accompanying materials. Its use is offered to you on the terms set out in this end-user license (the license). All rights not expressly granted are reserved to AAAS, OUP, or their suppliers.

As the publisher, OUP grants you, the buyer, the right to use one copy of the software on a single terminal connected to a single computer (a machine with a single CPU). You must treat the software and all accompanying materials like any other copyrighted materials (such as books and recordings). You may not reproduce the software in any form except that you may either make one copy for backup or archival purposes (provided you reproduce the copyright notice which appeared on the original media) or you may transfer the software to a single hard disk (provided you keep the original media solely for back-up or archival purposes). You may not transmit the software electronically to another machine. You may not alter, reverse engineer, decompile, or disassemble the software, and you may not copy the accompanying manual. You may not transfer any copy of the software to another user except that you may transfer the software and the accompanying materials to another user on a permanent basis provided that you retain no copies and the recipient agrees to the terms of this license.

You may not network the software or use it on more than one computer or terminal at the same time. You must obtain a supplementary license from OUP before using the software in connection with systems, multiple CPUs, or computer networks. Contact OUP at the address below for further information.

If you are provided with media in more than one format (such as both 3.5" and 5.25" diskettes or both Macintosh and IBM-compatible CD-ROM), you may use only one format. You may not use the other formats or loan, rent, lease, or transfer the other formats to another user.

You may decide to terminate this license by destroying the software and all copies thereof and all accompanying materials. If you violate the terms of this license in any way, the license is automatically terminated.

OUP warrants that the software and the accompanying materials shall be free from physical defects for a period of sixty days from the date of purchase. If notified at the address below of such defects during the warranty period, OUP will, at its sole discretion, replace the defective media or materials or refund your purchase price. The software and accompanying materials are provided "as is " and are not warranted to be free from errors or to operate without interruption. OUP and its suppliers assume no liability for any losses of any kind resulting from the use of this software, even if OUP has been in any way notified of the possibility of such damages. In no event will liability to OUP or its suppliers exceed the license fee you have paid to OUP.

This license is governed by the laws of the State of New York and the United States including patent and copyright laws. This license may not be modified except by express prior written agreement of the Product Manager, Electronic Publishing, OUP. If any provision of this license shall be found void or unenforceable, the balance of the license shall remain valid and enforceable. Use, duplication of, or disclosure of the software and its accompanying materials by the US Government shall be subject to the restricted rights applicable to commercial software under DFARS 52.227-7013.

Address questions comments or suggestions to AAAS, Project 2061, 1333 H Street, NW, P.O. Box 34446, Washington, DC 20005; phone 202-326-6666.; fax 202-842-5196.

For sales and marketing information, contact Oxford University Press, Department EC, 198 Madison Ave., New York, NY 10016: phone: 800-451-7556.